From Scratch To Numerical Methods: 25 Exercises

SVETLANA KARITSKAYA

From Scratch To Numerical Methods:

25 Exercises

Copyright © 2013 Svetlana Karitskaya

Copyright © 2013 Sergey Ponomarev

ISBN: 1490938702
ISBN-13: 978-1490938707

PREFACE

I like mathematics, I teach mathematics, I teach mathematics to students of various backgrounds, and I have observed over time that the vast majority of students do not like it. They do not like integrals, they do not like equations, and still, they need them. They need mathematics to succeed in the areas of their specializations, to get diplomas, to get a good job, and finally to realize that not liking mathematics just does not work in the long run. What does work in the long run is the knowledge. The knowledge of what to do, how to do, how to get it done. And because all of my students are (or going to be) engineers, they begin realizing (at some point in their lives) that mathematics is to be liked because...Ok, you know what they want to say, it is obvious. It was not obvious for them before, but now they understand that. And that's great! Along with them, I have realized that just teaching mathematics is not enough for students, and it is not enough for me either. If I like mathematics, and they don't, there is something wrong here. Not with them, but with me. And I have come to the realization that if I like it, and I want them to succeed in their journeys and become great engineers, it is not enough for me just to teach mathematics. I have to teach it in a way they would love it! That's how I came up with the idea of writing this book, specifically about numerical methods, to show that mathematics is not that difficult, it is fun, and everyone can enjoy it as well as I do.

To demonstrate this approach to my students, I wrote a short book about the most needed numerical applications for future engineers of all kinds of specialities, especially for those with a minimal knowledge of mathematics and who do need them to do their job.

Here they are, the three major applications of numerical methods, explained in the way that my students like!

1. Methods of calculations of definite integrals

2. Methods of solving differential equations

3. And methods of solving algebraic equations

I do believe that after reading through my book, most of the readers will get a better understanding of mathematics as a whole and as a useful and enjoyable discipline that every engineer needs to like and that they will like these particular numerical methods and find them practical in their activities.

Sincerely,

Svetlana Karitskaya

ACKNOWLEDGEMENTS

I am deeply grateful to all my colleagues at my work at the university and all my friends in Russia and USA for their support during and after the work on this book.

SVETLANA KARITSKAYA

CONTENTS

CHAPTER 1. Methods of approximate calculations of definite integrals

In solving problems of science and engineering by mathematical methods often there is a need to integrate a function.

It is known that every continuous function has a primitive. In such cases, a definite integral

$$\int_a^b f(x)dx \qquad (1)$$

of a continuous function $f(x)$ is usually calculated with the Newton-Leibniz formula. However, for a large class of elementary functions, primitive function is no longer elementary and cannot be defined by their combination. The following functions are examples of such functions:

$$\sqrt{x^5+1}\;;\; 1/\sqrt{x^3+1}\;;\; \frac{\sin x}{x}\;;\; \frac{\cos x}{x}\;;\; \sin(x^2)\;;\; \cos(x^2)\;;$$

$$\frac{e^x}{x^n}\;(n=1, 2, 3, ...)\;;\; e^{-x^2/2}\;;\; \frac{1}{\ln x}\;\text{ and there are many others.}$$

These functions cannot be integrated analytically, in other words, their primitive functions cannot be found. In those cases when the Newton-Leibniz formula is inapplicable, various methods of approximate calculation are used to calculate definite integrals.

The most common way of integrating any functions is a numerical integration, methods of which are mostly simple and easily translated into algorithmic languages (see Annex).

This book describes three most popular numerical methods.

1. The method of rectangles

If a function $f(x)$ is integrable in the interval $[a, b]$, then the definite integral is approximately equal to the following integral sum:

$$\int_a^b f(x)dx \approx \sum_{i=0}^{n-1} f(x_i)(x_{i+1} - x_i) \ \& \int_a^b f(x)dx \approx \sum_{i=1}^{n} f(x_i)(x_{i+1} - x_i) \ (2)$$

We divide the integration interval $[a, b]$ into n equal parts and denote the abscissas of the division points as

$$a = x_0 < x_1 < x_2 < ... < x_i < x_{i+1} < ... < x_n = b.$$

The length of each such a portion is $x_{i+1} - x_i = h = \dfrac{b-a}{n}$.

This value is called *step of integration.*

Values of the integrand $f(x)$ are calculated at each point of the division. In other words, the values are as follows:

$$f(x_0), \ f(x_1), \ f(x_2), \ ... , \ f(x_{n-1}).$$

We form the following integral sums:

$$\sum_{i=0}^{n-1} f(x_i)h = h\sum_{i=0}^{n-1} f(x_i) = \frac{b-a}{n}\left(f(x_0) + f(x_1) + f(x_2) + ... + f(x_{n-1})\right)$$

and

$$\sum_{i=1}^{n} f(x_i)h = h\sum_{i=1}^{n} f(x_i) = \frac{b-a}{n}\left(f(x_1) + f(x_2) + ... + f(x_{n-1}) + f(x_n)\right)$$

These expressions are the lower and upper integral sums, respectively.

The first one corresponds to the inscribed rectangles, and the second one corresponds to circumscribed rectangles. The constructed integral sums are the approximate values of the integral

$$\int_a^b f(x)dx \approx \frac{b-a}{n}\left(f(x_0)+f(x_1)+...+f(x_{n-1})\right) \qquad (3)$$

$$\text{or } \int_a^b f(x)dx \approx \frac{b-a}{n}\left(f(x_1)+f(x_2)+...+f(x_n)\right) \qquad (3')$$

Either of the resulting formulas (3) or (3') can be used to calculate the integral $\int_a^b f(x)dx$ by this method and it is called *formula of rectangles.*

Geometrically by the formula of rectangles, the area of a curvilinear trapezoid is approximately equal to the area of a step-like figure composed of rectangles (Fig. 1).

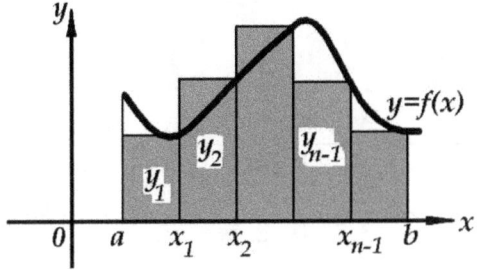

Fig.1

The error $\Delta(n)$ in calculating the value of the definite integral using the *formula of rectangles* is

$$\Delta(n) \le \frac{(b-a)^2}{n}|y'|, \qquad (4)$$

where $|y'|$ – the maximum of the absolute value of the derivative of the function $f(x)$ in the interval [a, b].

Example 1. Evaluate the integral $\int_0^4 (x^2 +1)dx$ by the Newton-Leibniz

3

and the approximate formula of rectangles, dividing the interval of integration into 8 equal pieces. Find the absolute and relative error of the calculation results by the approximate formula.

Solution. By the Newton-Leibniz formula

$$I = \int_0^4 (x^2 + 1)dx = \left(\frac{x^3}{3} + x\right)\Big|_0^4 = 25\frac{1}{3}.$$

The integration step is $h = \dfrac{4-0}{8} = 1/2 = 0{,}5$.

We make a table of values of the integrand $y = x^2 + 1$ for the division points obtained by dividing the integration interval [0, 4] into $n = 8$ equal parts:

i	0	1	2	3	4	5	6	7	$\sum\limits_{i=0}^{n-1} y_i$
x_i	0	1/2	1	3/2	2	5/2	3	7/2	
y_i	1	1,25	2	3,25	5	7,25	10	13,25	43

According to the formula of rectangles (3):

$$I_{rect} \approx h\sum_{i=0}^{7} y_i = 0{,}5(y_0 + y_1 + y_2 + y_3 + y_4 + y_5 + y_6 + y_7) = \frac{43}{2} = 21{,}5$$

The absolute error is $\Delta(n) = |I - I_{rect}| = 25\frac{1}{3} - 21\frac{1}{2} = \frac{23}{6} \approx 3{,}83$.

The relative error is $\delta = \dfrac{\Delta}{I} = \dfrac{23}{6} : 25\dfrac{1}{3} = \dfrac{23 \cdot 3}{6 \cdot 76} \approx 0{,}151 = 15{,}1\%$.

2. Euler's method (the trapezoidal rule)

Let the function be $f(x) \geq 0$, then the definite integral (1) can be interpreted as the area of the curvilinear trapezoid bounded by the graph of the function, the lines $x = a$, $x = b$ and $y = 0$ (Fig.2).

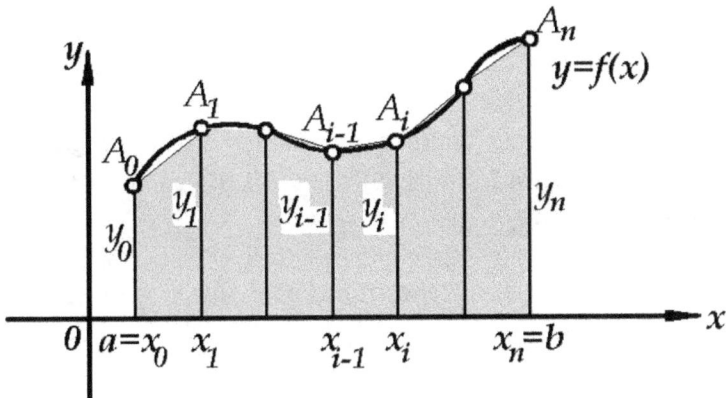

Fig. 2

We divide this interval $[a, b]$ into n equal parts using the same notation.

Through the points of division, draw lines parallel to the y-axis to the intersection with the curve. Connect the intersection points $A_0, A_1,..., A_n$ with line segments. Obviously, the area of the curvilinear trapezoid aA_0A_nb is approximately equal to the sum of squares n inscribed trapezoids whose bases are the coordinates y_i of the points A_i ($i = 0, 1, 2,$ ..., n), and the height of each segment is $h = \dfrac{b-a}{n}$.

Consequently, the area of the curvilinear trapezoid is

$$S = \int_a^b f(x)dx \approx \left(\frac{y_0 + y_1}{2} + \frac{y_1 + y_2}{2} + ... + \frac{y_{n-1} + y_n}{2} \right) h$$

or

$$\int_a^b f(x)dx \approx \frac{b-a}{n} \left(\frac{y_0 + y_n}{2} + y_1 + y_2 + ... + y_{n-1} \right) = h \left(\frac{y_0 + y_n}{2} + \sum_{i=1}^{n-1} y_i \right). \quad (5)$$

The equation (5) is called the *trapezoidal rule.* As it is shown below, the trapezoidal rule is more accurate than the formula of rectangles.

Note that the number n is chosen arbitrarily. Typically, the less the length of each interval, i.e. the greater the number of slots n, the less different the approximate and accurate values of the integral are, and the more accurately the formula (5) will yield the value of the integral. This is true for most functions.

The absolute error $\Delta(n)$ of the trapezoidal rule is expressed by the inequality

$$\Delta(n) \le \frac{(b-a)^3}{12n^2} |y''|, \quad (6)$$

where $|y''|$ – the maximum of the absolute value of the second derivative of the function $f(x)$ in the interval $[a, b]$. The proof of the inequality (6) is omitted.

Example 2. Evaluate the integral $\int_0^4 (x^2 +1)dx$ by the Newton-Leibniz formula and the approximate trapezoid rule, splitting the interval of integration into 8 equal pieces. Find the absolute and relative errors of the calculation results with the approximate formula. Compare the result with the previous example.

Solution. By the Newton-Leibniz formula:

$$I = \int_{0}^{4} (x^2 + 1)dx = \left(\frac{x^3}{3} + x \right)\Big|_{0}^{4} = 25\frac{1}{3} \approx 25{,}333 \, .$$

Now we find the approximate value of the integral by the trapezoidal rule. The integration step is $h = \dfrac{4-0}{8} = 1/2 = 0{,}5$.

We make a table of values of the integrand $y = x^2 + 1$ for the division points obtained by dividing the integration interval [0, 4] into $n = 8$ equal parts:

i	0	1	2	3	4	5	6	7	8	$\sum_{i=1}^{n-1} y_i$
x_i	0	1/2	1	3/2	2	5/2	3	7/2	4	
y_i	1	1,25	2	3,25	5	7,25	10	13,25	17	42

By the trapezoidal rule (5):

$$I_{TT} \approx h\left(\frac{y_0 + y_8}{2} + \sum_{i=1}^{7} y_i \right) = 0{,}5 \cdot \left(\frac{1+17}{2} + 42 \right) = \frac{51}{2} = 25\frac{1}{2} = 25{,}5 \, .$$

The absolute error is $\Delta(n) = |I - I_{TT}| = 25\dfrac{1}{2} - 21\dfrac{1}{3} = \dfrac{1}{6} \approx 0{,}167 \, .$

We estimate the absolute error by the formula (6). For this, we find the second derivative of the function $y = x^2 + 1$: $y' = 2x$, $y'' = 2$. Therefore, the largest absolute value of the second derivative of the function $y = x^2 + 1$ in the interval [0, 4] will be $|y''| = 2$ and

$$\Delta(n) \leq \frac{(b-a)^3}{12n^2}|y''| = \frac{(4-0)^3}{12 \cdot 8^2} 2 = \frac{1}{6} \approx 0{,}167 \, .$$

The relative error is:

$$\delta = \frac{\Delta}{I} = \frac{1}{6} : 25\frac{1}{3} = \frac{1 \cdot 3}{6 \cdot 76} = \frac{1}{152} \approx 0,007 = 0,7\,\%.$$

Obviously, the accuracy of the calculations using the formula of rectangles is clearly insufficient. When using the trapezoidal rule, we have come much closer to the desired value.

3. Simpson's method (parabolas formula)

According to Euler's method (the trapezoidal rule), the area of the curvilinear trapezoid is calculated in each interval. If we combine the two intervals, the area under the graph of the function $f(x)$ over two intervals can be approximated not with an area of two trapezoids, but the area under the parabola on the double interval (see Fig.3). This approach is called *Simpson's method or rule (parabolas formula)*. We adopt this method without a rigorous proof.

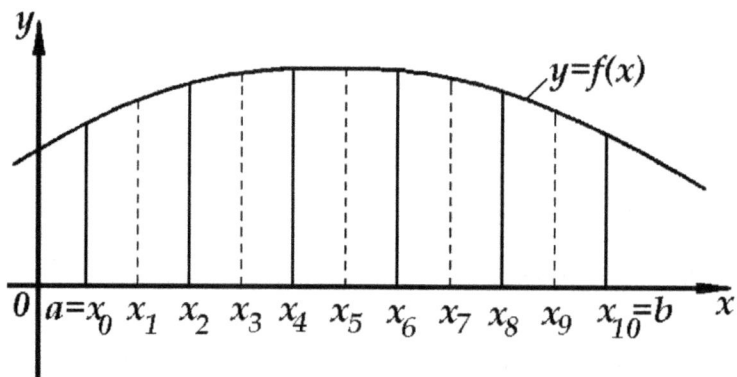

Fig. 3

In the example shown in Fig. 3, the interval [a, b] is divided into 10 intervals. The length h of one interval is $h = \dfrac{b-a}{n} = \dfrac{x_n - x_o}{n}$, where n – the number of the intervals.

If we assume that the curve on the interval $[x_0, x_2]$ is parabola, then the area S under it can be accurately calculated by the formula

$$S = \frac{h}{3}(f(x_0) + 4f(x_1) + f(x_2))$$

If we combine all the intervals, we get the following formula:

$$S = \frac{h}{3} \: [f(x_0) + 4f(x_1) + 2f(x_2) + 4f(x_3) + 2f(x_4) + 4f(x_5) + 2f(x_6) +$$
$$+ \: 4f(x_7) + 2f(x_8) + 4f(x_9) + f(x_{10})].$$

Let's have an arbitrary function $f(x)$; it is necessary to calculate the definite integral of the function in the range from a to b.

When integrating by Simpson's method, it is necessary to divide the integration interval $[a, b]$ into an even number of equal intervals. We denote this number by $2n$, the division points of the segment $[a, b]$ as

$$a = x_0 < x_1 < x_2 < ... < x_{2n-2} < x_{2n-1} < x_{2n} = b,$$

and the values of the integrand at these points as

$$y_0, \quad y_1, \quad y_2, \quad ...., \quad y_{2n-2}, \quad y_{2n-1}, \quad y_{2n} .$$

The formula for the approximate calculation of the definite integral in this case is written as follows:

$$\int_a^b f(x)dx \approx \frac{b-a}{6n}\left(y_0 + y_{2n}\right) + 4(y_1 + y_3 + y_5 + ... + y_{2n-1}) +$$

$$+ \: 2(y_2 + y_4 + y_6 + ... + y_{2n-2})\Big).$$

Or, using the notations

$$y_0 + y_{2n} = \Sigma_1,$$

$$y_1 + y_3 + y_5 + \ldots + y_{2n-1} = \sum_{i=1}^{n} y_{2i-1} = \Sigma_2$$

and

$$y_2 + y_4 + y_6 + \ldots + y_{2n-2} = \sum_{i=2}^{n} y_{2i-2} = \Sigma_3,$$

we write the formula for the approximate calculation of integrals, called *parabolas formula* or *Simpson's formula:*

$$\int_a^b f(x)dx \approx \frac{b-a}{6n}\left((y_0 + y_{2n}) + 4\sum_{i=1}^{n} y_{2i-1} + 2\sum_{i=2}^{n} y_{2i-2}\right); \quad (7)$$

or $$\int_a^b f(x)dx \approx \frac{b-a}{6n}(\Sigma_1 + 4\Sigma_2 + 2\Sigma_3). \quad (8)$$

The absolute error $\Delta(n)$ *of Simpson's formula* is expressed by the inequality

$$\Delta(n) \le \frac{(b-a)^5}{180n^4}\left|y^{IV}\right|,$$

where $\left|y^{IV}\right|$ is the maximum absolute value of the forth-order derivative of the function $f(x)$ in the interval [a, b]. The error that occurs when using the approximate method is reduced in proportion to the length of the interval to the fourth power, that is, when the number of intervals n increases twice, the calculation errors $\Delta(n) \approx h^4$ decreases 16 times.

Example 3. Evaluate the integral $\int_0^4 (x^2 + 1)dx$ by the approximate

10

Simpson's formula, breaking the interval of integration into 8 equal pieces.

Solution. In this method, the number of parts of the division should be considered equal to $2n = 8$, hence $n = 4$.

We calculate the factor standing in front of the bracket in the formula (8):

$$\frac{b-a}{6n} = \frac{4-0}{6 \cdot 4} = \frac{1}{6}.$$

We make a table of values of the integrand $y = x^2 + 1$ for the division points obtained by dividing the integration interval [0, 4] into $2n = 8$ equal parts:

i	0	1	2	3	4	5	6	7	8
x_i	0	1/2	1	3/2	2	5/2	3	7/2	4
y_i	1	1,25	2	3,25	5	7,25	10	13,25	17
$\Sigma_1 = y_0 + y_8$					18				
$\Sigma_2 = y_1 + y_3 + y_5 + y_7$					25				
$\Sigma_3 = y_2 + y_4 + y_6$					17				

According to Simpson's formula (8):

$$I_{Simp} \approx \frac{b-a}{6n}\left(\Sigma_1 + 4\Sigma_2 + 2\Sigma_3\right) = \frac{1}{6} \cdot (18 + 4 \cdot 25 + 2 \cdot 17) = \frac{152}{6} = 25\frac{1}{3} = 25{,}33.$$

The absolute error is: $\Delta(n) = \left| I - I_{\text{Simp}} \right| = 25\frac{1}{3} - 25\frac{1}{3} = 0$.

Example 4. Calculate the number π by using the integral

$\int_0^1 \frac{dx}{x^2+1} = \text{arctg } x \big|_0^1 = \frac{\pi}{4}$. (Compare the result with the number

$\frac{\pi}{4} \approx 0{,}785\,398$.)

Solution. The integration interval is divided into 10 parts. The calculations are performed in three ways.

The method of rectangles. The integration step is $h = \frac{1-0}{10} = 0{,}1$.

We make a table of values of the integrand $y = \frac{1}{x^2+1}$.

(see next page)

i	x_i	$y_i = \dfrac{1}{x_i^2 + 1}$	$\displaystyle\sum_{i=0}^{n-1} y_i$
0	0,0	$\dfrac{1}{0^2 + 1} = \dfrac{1}{1} = 1{,}000\,000$	
1	0,1	$\dfrac{1}{0{,}1^2 + 1} = \dfrac{1}{1{,}01} = 0{,}990\,099$	
2	0,2	$\dfrac{1}{0{,}2^2 + 1} = \dfrac{1}{1{,}04} = 0{,}961\,538$	
3	0,3	$\dfrac{1}{0{,}3^2 + 1} = \dfrac{1}{1{,}09} = 0{,}917\,431$	
4	0,4	$\dfrac{1}{0{,}4^2 + 1} = \dfrac{1}{1{,}16} = 0{,}862\,069$	
5	0,5	$\dfrac{1}{0{,}5^2 + 1} = \dfrac{1}{1{,}25} = 0{,}800\,000$	8,099 814
6	0,6	$\dfrac{1}{0{,}6^2 + 1} = \dfrac{1}{1{,}36} = 0{,}735\,294$	
7	0,7	$\dfrac{1}{0{,}7^2 + 1} = \dfrac{1}{1{,}49} = 0{,}671\,141$	
8	0,8	$\dfrac{1}{0{,}8^2 + 1} = \dfrac{1}{1{,}64} = 0{,}609\,756$	
9	0,9	$\dfrac{1}{0{,}9^2 + 1} = \dfrac{1}{1{,}81} = 0{,}552\,486$	

According to the formula of rectangles (3):

$$I_{rect} \approx h\sum_{i=0}^{9} y_i = 0{,}1(y_0 + y_1 + y_2 + y_3 + y_4 + y_5 + y_6 + y_7 + y_8 + y_9) =$$

$$= 0{,}809\,981$$

Comparing with the proposed value, we see that the accuracy obtained is low.

The absolute error of the calculation of the number π by the formula of rectangles is:

$$\Delta(n) = \left| \frac{\pi}{4} - I_{rect} \right| = |0{,}785\,398 - 0{,}809\,981| = 0{,}024\,583 .$$

The relative error is: $\delta = \dfrac{0{,}024\,583}{0{,}785\,398} \approx 0{,}031\,301 = 3{,}1\,\% .$

The trapezoidal rule

The integration step is $h = \dfrac{1-0}{10} = 0{,}1$.

We make a table of values of the integrand $y = x^2 + 1$.

i	x_i	$y_i = \dfrac{1}{x_i^2 + 1}$	$y_0 + y_n$	$\displaystyle\sum_{i=1}^{n-1} y_i$
0	0,0	1,000 000	1,5	7,099 814
1	0,1	0,990 099		
2	0,2	0,961 538		
3	0,3	0,917 431		
4	0,4	0,862 069		
5	0,5	0,800 000		
6	0,6	0,735 294		
7	0,7	0,671 141		
8	0,8	0,609 756		
9	0,9	0,552 486		
10	1	0,500 000		

According to the trapezoidal rule (5):

15

$$I_{\text{тр}} \approx h\left(\frac{y_0 + y_{10}}{2} + \sum_{i=1}^{9} y_i\right) = 0,1 \cdot \left(\frac{1,5}{2} + 7,099\,814\right) \approx 0,784\,981.$$

The absolute error of the calculation of the number $\pi/4$ by the trapezoidal rule is:

$$\Delta(n) = \left|\frac{\pi}{4} - I_{\text{тр}}\right| = |0,785\,398 - 0,784\,981| \approx 0,000\,417.$$

This result indicates that the desired value obtained by the trapezoidal rule coincides with the proposed value up to the third digit.

The relative error is:

$$\delta = 0,004\,170 : 0,785\,398 \approx 0,000\,530 = 0,5\,\%.$$

Simpson's formula

The number of parts of the division is *2n* = 10, hence *n* = 5.

We calculate the factor standing in front of the bracket in the formula (8):

$$\int_a^b f(x)dx \approx \frac{b-a}{6n}(\Sigma_1 + 4\Sigma_2 + 2\Sigma_3),$$

$$\frac{b-a}{6n} = \frac{1-0}{6 \cdot 5} = \frac{1}{30}.$$

We make a table:

i	x_i	$y_i = \dfrac{1}{x_i^2 + 1}$
0	0	1,000 000
1	0,1	0,990 099
2	0,2	0,961 538
3	0,3	0,917 431
4	0,4	0,862 069
5	0,5	0,800 000
6	0,6	0,735 294
7	0,7	0,671 141
8	0,8	0,609 756
9	0,9	0,552 486
10	1,0	0,500 000
$\Sigma_1 = y_0 + y_{10}$		1,5
$\Sigma_2 = y_1 + y_3 + y_5 + y_7 + y_9$		3,931 157
$\Sigma_3 = y_2 + y_4 + y_6 + y_8$		3,168 657

According to Simpson's formula (8):

$$I_{\text{Simp}} \approx \frac{b-a}{6n}\left(\Sigma_1 + 4\Sigma_2 + 2\Sigma_3\right) = \frac{1}{30}\cdot\left(1,5 + 4\cdot 3,931\,157 + 2\cdot 3,168\,657\right)=$$
$$= 0,785\,398.$$

The absolute error is::

$$\Delta(n) = \left| \frac{\pi}{4} - I_{Simp} \right| = \left| 0,785\,398 - 0,785\,398 \right| = 0, \text{ i.e. by this method,}$$

the value of the number $\frac{\pi}{4}$ was obtained with accuracy up to six characters.

Exercises

1. Calculate the following integrals by the Newton-Leibniz and the approximate formulas of rectangles, trapezoids and parabolas.

1.1. $\displaystyle\int_{-1}^{5} \left(x^2 - 2 \right) dx$ $(n = 6)$. **1.2.** $\displaystyle\int_{0}^{4} \sqrt{x}\,dx$ $(n = 8)$. **1.3.** $\displaystyle\int_{0}^{1} e^x dx$ $(n = 8)$.

1.4. $\displaystyle\int_{0}^{\pi/2} \sin x\,dx$ $(n = 6)$. **1.5.** $\displaystyle\int_{0}^{0,8} \cos x\,dx$ $(n = 10)$. **1.6.** $\displaystyle\int_{0,1}^{1,7} \frac{dx}{\sqrt{x}}$ $(n = 8)$.

2. Calculate the following integrals (that cannot be found in the final form using elementary functions) by approximate trapezoidal and Simpson's formulas.

2.1. $\displaystyle\int_{0}^{3} \sqrt{x^3 + 1} \cdot dx$ $(n = 6)$.   **2.2.** $\displaystyle\int_{0}^{\pi/3} \sqrt{\cos x} \cdot dx$ $(n = 10)$.

2.3. $\displaystyle\int_{0}^{1} \frac{\operatorname{arctg} x}{x} dx$ $(n = 10)$.   **2.4.** $\displaystyle\int_{0}^{1} e^{-x^2} dx$ $(n = 10)$.

Note: For $x = 0$, the value $\dfrac{\operatorname{arctg} x}{x}$ is determined as $\displaystyle\lim_{x \to 0} \dfrac{\operatorname{arctg} x}{x} = 1$.

CHAPTER 2. Approximate calculations of the solutions of differential equations of the first order $y' = f(x, y)$ by Euler's method

In solving scientific and engineering problems is often necessary to mathematically describe a dynamic system. This is best done in the most natural form - in the form of differential equations or systems of differential equations. Here are a few examples:

a)   The kinetics of the chemical reactions – $c' = f(c, t)$ (c can mean concentration of the substance as well as the vector of concentrations);

b)   Transport phenomena (transfer of heat, mass and momentum). Under steady-state conditions for the description of these processes one can use ordinary differential equations; at non-stationary conditions, these phenomena are described by partial differential equations. In engineering, processes such as heat and mass transfer are described by the corresponding equations that are empirical (e.g., in the case of heat transfer, stirring, drying, extraction, adsorption, and many others);

c)   Population dynamics. Changes in populations of predators and their preys can be described, for example, by two relatively simple differential equations, assuming that the dependence of birth and death of individual animals from their "concentration" is known;

d)   The spread of "new." The spread of epidemics, rumors, opinions and experience in certain conditions, can also be described by differential equations. For epidemics, one should consider the differences between the "healthy", "positive", "resistant" and "isolated" populations. Kinetics of each of these populations, in accordance with the selected model, may vary.

In solving differential equations one can meet three types of the Cauchy problem:

a) Differential equations *with given initial conditions.* This problem occurs, for example, when solving differential equations that describe the kinetics of chemical reactions. Concentrations of the substances at the initial time *(t* = 0) are generally known and it is necessary to find the concentration of substances after a certain period of time *t;*

b) The *boundary value problem.* In this case, we know the value of the function or its derivatives at certain points and it is necessary to find a solution to the differential equation between these points. An example of the boundary value problem is the calculation of beam deflection under load. If the ends of the beams are fixed, then both ends of the beam deflection is zero. Another example - a stationary temperature distribution in a cube, three faces of which have the temperature 20 ° C, and the other three faces have the temperature 80 ° C;

c) The *eigenvalue problem.* Problems of this type are very similar to the boundary value problem. Let me explain this with the following example. In the Schrödinger equation for the harmonic oscillator, the total energy of the system is included as a parameter. We are interested in for which values of energy Schrödinger equation has "reasonable" solutions with respect to the physical content of the problem. These solutions must correspond to such wave functions which decrease to zero with increasing distance. The energy values that satisfy this condition are called eigenvalues, and the corresponding wave functions - private functions. Another example of a simple eigenvalue problem - the differential equation describing the vibration of the string. The boundary conditions in this case are defined as the string is fixed at both ends. In the differential equation describing the vibrations of a string there is a parameter that corresponds to the wavelength. We need to find what values it can take, that is, for what values of the parameter the differential equation satisfies the boundary conditions.

However, only in rare cases it is possible to obtain a function in an analytical form that satisfies the given equation and the initial condition. If none of the particular methods of integration of the equation of the first order

$$y' = f(x, y) \qquad (9)$$

leads to the goal or require complex calculations, then you can go for an approximate solution. Here we present a graphical method of Euler and, arising from it, the method of numerical integration.

Before we turn to the consideration of these methods, let's touch on the geometric meaning of the first order equation (9).

1. Geometric meaning of a first order differential equation

At each point P(x, y) of the region of the x-y plane where the theorem of existence and uniqueness is valid, the equation (9) determines the value of the slope of the tangent (y') to the integral curve passing through the point P(x, y). This value can be represented graphically by a straight arrow, starting at the point P and having a slope equal to f (x, y); the length of the arrow is irrelevant. Thus, having set the equation (9), field lines in the xy-plane are set as well.

The locus of points in the same direction of the field ($y' = \mathrm{const}$) is called *isocline* (line of equal slope) equation. Obviously, we obtain the equation of the isocline corresponding to a given value $y' = C$, if we substitute this value in the differential equation

$$C = f(x, y).$$

For arbitrary but fixed C the latter is the equation of the family of isoclines of the differential equation (9).

At all points of one isocline corresponding to one C, tangents to the integral curves have the same direction.

Thus, the task of integrating a differential equation can be geometrically interpreted as follows: *find the lines that satisfy the condition that the tangents to them have the directions coinciding with the direction of the field at the points of contact.*

Let's proceed to the description of the methods of approximate integration, and we assume that the theorem of existence and uniqueness of solution is valid.

2. Euler's graphical method

Consider the Euler's method of constructing an integral curve of the first order

$$y' = f(x, y),$$

passing through the initial point $M_0(x_0, y_0)$ i.e. a graphical finding a particular solution with the initial condition

$$y\Big|_{x=x_0} = y_0.$$

Roughly, it can be done with simple constructions that are similar to those produced in the graphical integration of functions, i.e. solving the equation (9) in the special case, when $f(x, y)$ is a function of x only.

Divide the interval $[x_0, x]$ into n parts by the points $x_1, x_2, \ldots x_{n-1}$ (Fig. 4). Draw lines parallel to the *y-axis* through the points of division, and consequentially we carry out the following the same operations.

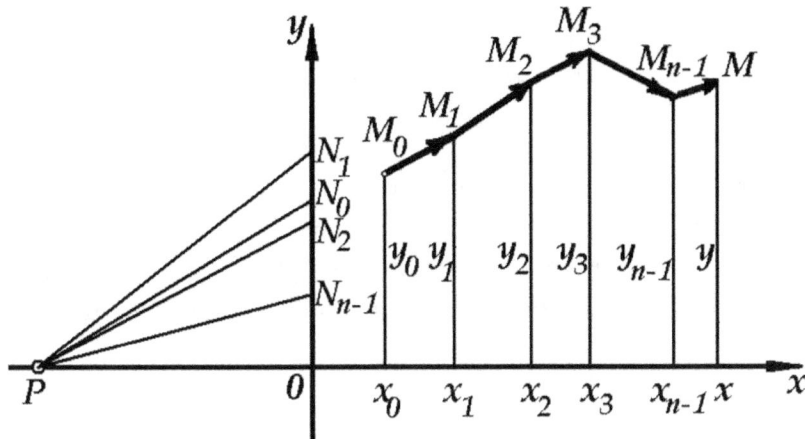

Fig. 4

We calculate the value $f(x,y)$ at $M_0(x_0, y_0)$; $f(x_0, y_0)$ measures, according to equation (9), the angular coefficient of the integral curve at the point $M_0(x_0, y_0)$. To draw the direction coming from M_0, take the pole P of the chart located at the distance $OP = 1$ on the x-axis to the left from the beginning of the coordinates (the scale of the OP may be different from the scale taken along the coordinate axes); draw the segment ON_0, equal to $f(x_0, y_0)$ on the scale OP, on the axis Y-axis and connect the point N_0 with the pole P by means of a straight line. The direction of the segment PN_0 obviously will be the desired direction of the curve at the point M_0. Then, we draw a line from M_0, parallel to PN_0, to its intersection with the line $x = x_1$. We get the point M_1 which will be the point of the integral curve corresponding to $x = x_1$. This construction means replacing the arc in the partial interval $[x_0, x_1]$ with a segment of its tangent at the starting point.

Next, we calculate the value of $f(x,y)$ found in the point $M_1(x_1, y_1)$; the value $f(x_1, y_1)$ characterizes the slope of the curve at M_1. On the axis Y-axis , we make the segment ON_1, equal to $f(x_1, y_1)$ on the scale OP, and connect the point N_1 with the pole P. Then, from the point M_1, we draw a line segment, parallel to PN_1, to the intersection with the line $x = x_2$. At the intersection we get the point M_2, which is taken as a point of the integral curve corresponding to $x = x_2$.

In the same way, we find one by another points of the curve corresponding to the points of the partition x_3, x_4, ..., of the interval $[x_0,$ $x_1]$, until we reach the point $M(x, y)$. The drawn broken line $M_0M_1M_2...M_{n-1}M$ is approximately the integral curve passing through the point $M_0(x_0, y_0)$.

3. Numerical integration

Let's translate Euler's method of approximate integration of the differential equation (9) into analytical language.

We divide the interval $[x_0, x]$ with the points x_0, x_1, x_2, ..., x_{n-1}, $x_n = x$ into n equal parts (where $x_0 < x_1 < x_2 < ... < x_{n-1} < x_n$). We denote $x_1 - x_0 = x_2 - x_1 = ... = x - x_{n-1} = \Delta x$. Then

$$\Delta x = \frac{x - x_0}{n}.$$

Let $y = \varphi(x)$ denote an approximate solution of the equation (9) and

$$y_0 = \varphi(x_0), \ y_1 = \varphi(x_1), \ ..., \ y_n = \varphi_n(x).$$

We denote

$$\Delta y_0 = y_1 - y_0, \ \Delta y_1 = y_2 - y_1, \ ..., \ \Delta y_{n-1} = y_n - y_{n-1}.$$

In each of the points x_0, x_1, x_2, ..., x_n in the equation (9), we replace the derivatives with the ratio of the finite differences:

$$\frac{\Delta y}{\Delta x} = f(x, y), \tag{10}$$

$$\Delta y = f(x, y)\Delta x. \tag{10'}$$

At $x = x_0$ we have

$$\frac{\Delta y_0}{\Delta x} = f(x_0, y_0), \ \Delta y_0 = f(x_0, y_0)\Delta x.$$

Thus, the first operation gives the following relationship between the coordinates of the points M_0 и M_1:

$$y_1 - y_0 = f(x_0, y_0) \cdot (x_1 - x_0) = f(x_0, y_0) \cdot \Delta x. \tag{11}$$

In this equation $x_0, y_0, \Delta x$ are known. Consequently, we find

$$y_1 = y_0 + f(x_0, y_0)\Delta x.$$

At $x = x_1$ the equation (10') takes the form

$$\Delta y_1 = f(x_1, y_1)\Delta x.$$

Thus, the second operation results in a similar relationship:

$$y_2 - y_1 = f(x_1, y_1) \cdot (x_2 - x_1); \ y_2 = y_1 + f(x_1, y_1)\Delta x. \tag{12}$$

Similarly, we find

$$y_3 = y_2 + f(x_2, y_2)\Delta x \tag{13}$$

etc., and finally, *n-th* operation gives

$$y - y_{n-1} = f(x_{n-1}, y_{n-1}) \cdot (x - x_{n-1}); \ y = y_{n-1} + f(x_{n-1}, y_{n-1}) \cdot \Delta x. \tag{14}$$

These *n* equations can consistently calculate the values of the unknown function at the points of division of the interval $[x_0, x]$. Indeed, from the first equation for given x_0, y_0 and selected x_1 we find y_1, from the second one with the known x_1, y_1 and selected x_2, we find the corresponding value y_2 and so on, until we reach the required value of *y*.

The smaller the greatest of the partial intervals, and hence, the greater the number n and the closer x to x_0, the more accurate the result will be.

Thus, the actual calculation of the approximate value of y can be done with the help of the equations (11)–(14). This method is a *method of numerical integration* of the equation (9). According to the method, the calculation error is proportional to the step's length Δx.

By connecting the points (x_0, y_0), (x_1, y_1), ..., (x_n, y_n) on the coordinate plane with line segments, we get a broken line – an approximate representation of the integral curve (see Fig.4). This broken line is called *the Euler line.*

Remark. Denote an approximate solution of the equation (9) as $y = \varphi_h(x)$ which corresponds to the Euler line at Δx. We can prove that if there exists a unique solution $y = \varphi^*(x)$ of the equation (9), satisfying the initial conditions and defined in the interval $[x_0, x]$, then $\lim\limits_{\Delta x \to 0} |\varphi_h(x) - \varphi^*(x)| = 0$ for each x in the interval $[x_0, x]$.

Example 1. Consider the equation $y' = xy^2+1$. None of the analytical methods of solving equations of the first order can be used for this one. We find an approximate solution of this equation on the interval $[0, 1]$ with the initial condition $y\big|_{x=0} = 0$ and compute y when $x = 1$.

We divide the interval $[0, 1]$ into four parts with the points $x_0 = 0$; $x_1 = 0{,}25$; $x_2 = 0{,}5$; $x_3 = 0{,}75$; $x_4 = 1$ (Fig. 5).

We denote the values of y' as y_0, y_1, y_2, y_3 in the points x_0, x_1, x_2, x_3, respectively.

Since $x_0 = 0$, $y_0 = 0$, then $y_0' = 1$ and, therefore,

$$y_1 = 1 \cdot (x_1 - x_0) + 0 = 0{,}25 \, ; \quad M_1 \, (0{,}25; \, 0{,}25).$$

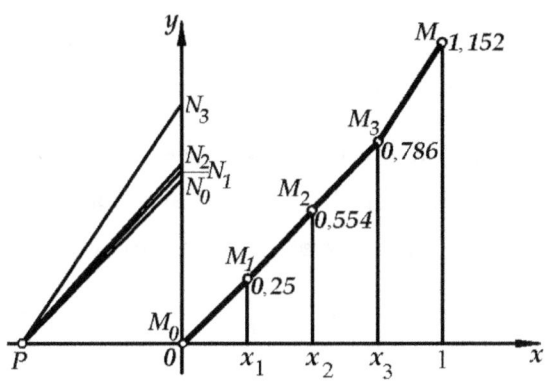

Fig. 5

Next

$$y_1' = 0{,}25 \cdot 0{,}25^2 + 1 = 1{,}016 ,$$

and that's why

$$y_2 = 1{,}016 \cdot 0{,}25 + 0{,}25 = 0{,}504 ; \quad M_2(0{,}5; 0{,}504).$$

Then

$$y_2' = 0{,}5 \cdot 0{,}504^2 + 1 = 1{,}127 ,$$

$$y_3 = 1{,}127 \cdot 0{,}25 + 0{,}504 = 0{,}786 ; \quad M_3(0{,}75; 0{,}786).$$

Finally,

$$y_3' = 0{,}75 \cdot 0{,}786^2 + 1 = 1{,}463$$

and

$$y = y_4 = 1{,}463 \cdot 0{,}25 + 0{,}786 = 1{,}152 .$$

Therefore, $y = 1{,}152$ is the desired approximate value for $x = 1$ of the particular solution of the given equation, defined by $y|_{x=0} = 0$.

In Fig. 5, the integral curve of the equation passing through the point (0; 0) is plotted in accordance with the calculations.

Example 2. Find an approximate (at $x = 1$) value of the solution of the equation $y' = y + x$, satisfying the initial condition $y|_{x=0} = 1$.

Solution. We divide the interval $[0;1]$ into 10 equal parts. The step is $\Delta x = \dfrac{1-0}{10} = 0{,}1$. The values of y_i ($i = 1,2,...,10$) will be calculated by the formula (10'):

$$\Delta y_i = (y_i + x_i)\Delta x$$

or

$$y_{i+1} = y_i + (y_i + x_i)\Delta x.$$

The calculations are tabulated below.

i	x_i	y_i	$y_i + x_i$	$\Delta y_i = (y_i + x_i)\Delta x$
0	0	1,0000	1,0000	0,1000
1	0,1	1,1000	1,2000	0,1200
2	0,2	1,2200	1,4200	0,1420
3	0,3	1,3620	1,6200	0,1620
4	0,4	1,5240	1,9240	0,1924
5	0,5	1,7164	2,2164	0,2216
6	0,6	1,9380	2,5380	0,2538
7	0,7	2,1918	2,8918	0,2892
8	0,8	2,4810	3,2810	0,3281
9	0,9	2,8091	3,7091	0,3709
10	1,0	3,1800		

The approximate value found is as follows: $y\big|_{x=1} = 3,1800$.

The analytical solution of this equation that satisfies the specified initial condition has the form:

$$y = 2e^x - x - 1.$$

Therefore, a more accurate solution of the equation

$$y\big|_{x=1} = 2(e-1) \approx 3,4366.$$

The absolute error of the numerical solution is: $\Delta = 0,2566$;

and the relative error is $\delta = \dfrac{0{,}2566}{3{,}4366} = 0{,}075 \approx 8\,\%$.

Exercises

1. Solve numerically the following differential equations for the initial condition $y\big|_{x=0} = 1$. Find solutions for several values of the independent variable and create a chart. Compare the results of the numerical and analytical methods (in cases where the equation can be solved analytically.)

1.1. $y' = -y$.

1.2. $y' = k - y^2$.

1.3. $y' = x^y + y^x$.

1.4. $y' = -ky^2 + \sin^2(xy)$. **1.5.** $y' = 2y + e^x$.

1.6. $y' = x^3 + 2xy$.

1.7. $y' = \ln x + \dfrac{x}{y}$.

1.8. $y' = -0{,}1y - 0{,}5y^2$. **1.9.** $y' = 1 - y - e^{-x}$.

1.10. $y' = \sin(xy) + \cos\dfrac{x}{y}$. **1.11.** $y' = y\ln x - y$. **1.12.** $y' = xy^3 - y$.

1.13. $y' = e^x y^2 - 2y$.

1.14. $y' = xy$.

1.15. $y' = \ln y$.

2. Solve these differential equations with other (arbitrary) initial conditions.

3. Write a computer program to solve one of these equations.

CHAPTER 3. Approximate calculations of definite integrals and finding particular solutions of differential equations using power series

Power series have a variety of applications. They can be used to calculate the values of functions with any degree of accuracy, to solve uncertain and definite integrals, integration of differential equations.

Before discussing the application of power series to approximate calculations, we illustrate the expansion of functions e^x, $\sin x$, $\cos x$, $\ln(1+x)$ and $(1+x)^m$ in a Maclaurin series.

1. Presentation of some elementary functions as a Maclaurin series

1.1. $e^x = 1 + \dfrac{x}{1!} + \dfrac{x^2}{2!} + \dfrac{x^3}{3!} + \dots + \dfrac{x^n}{n!} + \dots,\ x \in (-\infty; +\infty).$ (15)

1.2. $\sin x = x - \dfrac{x^3}{3!} + \dfrac{x^5}{5!} - \dfrac{x^7}{7!} + \dots + (-1)^n \dfrac{x^{2n+1}}{(2n+1)!} + \dots, x \in (-\infty; +\infty)$ (16)

1.3. $\cos x = 1 - \dfrac{x^2}{2!} + \dfrac{x^4}{4!} - \dfrac{x^6}{6!} + \dots + (-1)^n \dfrac{x^{2n}}{(2n)!} + \dots, x \in (-\infty; +\infty)$ (17)

1.4. $\ln(1+x) = x - \dfrac{x^2}{2} + \dfrac{x^3}{3} - \dfrac{x^4}{4} + \dots + (-1)^{n+1} \dfrac{x^n}{n} + \dots, x \in (-1; +1]$ (18)

1.5. $(1+x)^m = 1 + mx + \dfrac{m(m-1)}{2!}x^2 + \dots + \dfrac{m(m-1)(m-2)\dots(m-n+1)}{n!}x^n + \dots x \in (-1; +1)$ (19)

where m is any real number.

1.6. $\dfrac{1}{(1+x)} = 1 - x + x^2 - x^3 + ... + (-1)^n x^n + ... \; x \in (-1; +1)$ (20)

1.7. $\operatorname{arctg} x = x - \dfrac{x^3}{3} + ... + (-1)^{n+1} \dfrac{x^{2n+1}}{2n+1} + ... \; x \in (-1; +1)$ (21)

2. Application of power series to calculate definite integrals

In integral calculus, it is known that many of the integrals cannot be expressed in a closed-form in terms of elementary functions. In such cases, an integral can be calculated approximately by the methods outlined above. Another way to calculate these integrals is the expansion of the integrand in a power series and its integration term by term on the basis of the following theorem.

The theorem on the integration of power series. If a power series

$$a_0 + a_1 x + a_2 x^2 + a_3 x^3 + ... + a_n x^n + ...$$

converges on the interval $(-R; +R)$ and the function $f(x)$ is its sum, then the power series obtained from its integration term by term has the same interval of convergence, and its sum is that of the primitive functions of $f(x)$ which is equal to zero when $x = 0$.

Let's explain this method with examples.

Example 1. Calculate the integral $\int\limits_0^1 \sqrt{x}\,e^{-x}\,dx$.

Solution. "Exact" integration is impossible because this integral refers to non integrable. By replacing x with $(-x)$ in (15), we obtain:

$$e^{-x} = 1 - \frac{x}{1!} + \frac{x^2}{2!} - \frac{x^3}{3!} + \ldots + (-1)^n \frac{x^n}{n!} + \ldots .$$

After multiplying the resulting power series by $\sqrt{x}$

$$\sqrt{x}\,e^{-x} = x^{1/2}e^{-x} = x^{1/2} - \frac{x^{3/2}}{1!} + \frac{x^{5/2}}{2!} - \frac{x^{7/2}}{3!} + \ldots + (-1)^n \frac{x^{n+1/2}}{n!} + \ldots$$

and integrating it term by term in the interval $[0;1]$, belonging to the interval of convergence $(-\infty; +\infty)$, we obtain

$$\int\limits_0^1 \sqrt{x}\,e^{-x}\,dx = \int\limits_0^1 x^{1/2}\,dx - \int\limits_0^1 x^{3/2}\,dx + \ldots + \int\limits_0^1 (-1)^n \frac{x^{n+1/2}}{n!}\,dx + \ldots =$$

$$= \frac{2}{3}x^{3/2}\Big|_0^1 - \frac{2}{5}x^{5/2}\Big|_0^1 + \ldots + \frac{(-1)^n}{n!} \cdot \frac{x^{n+3/2}}{n+3/2}\Big|_0^1 =$$

$$= 0,66\,667 - 0,40\,000 + 0,14\,286 - 0,03\,704 + 0,00\,758 - 0,00\,128 +$$

$$+ 0,00\,018 - \ldots \approx 0,37\,897 .$$

We take the first six terms of the expansion, based on the corollary of the Leibnitz rule for convergent alternating series, and we let the error not to exceed the first discarded term in absolute value, i.e. $\Delta \le 0,00\,018$.

As the seventh discarded member is positive, the value of the integral was found insufficiently.

Example 2. Calculate the integral $\int\limits_0^1 \sqrt{x}e^x dx$ to the nearest 0,0001.

Solution. The expression of this integral in the form of a numerical series is similar to the previous example:

$$\int\limits_0^1 \sqrt{x}e^x dx = \int\limits_0^1 x^{1/2} dx + \int\limits_0^1 x^{3/2} dx + ... + \int\limits_0^1 \frac{x^{n+1/2}}{n!} dx + ... =$$

$$= \frac{2}{3} + \frac{2}{5} + \frac{2}{7 \cdot 2!} + \frac{2}{9 \cdot 3!} + ... + \frac{2}{(2n+3)n!}.$$

In contrast, the calculation of the integral does not come down to finding the sum of a convergent alternating series, the error in the calculation of which is estimated by the corollary of the Leibniz rule, but to determine the sum of the series with positive terms with the unknown error estimate.

We proceed as follows. Let's suppose that to estimate the sum of the series we have taken a number n of the series terms (along with the first one at $n = 0$). Then the error of calculation will be determined by the sum of a remainder of the series:

$$\Delta = r_n = \frac{2}{(2n+3)n!} + \frac{2}{(2n+5)(n+1)!} + \frac{2}{(2n+7)(n+2)!} + ...$$

As

$$2n+5 > 2n+3, \ 2n+7 > 2n+3, \ ...$$

and

$$(n+1)! = n!(n+1) > n!n, \ (n+2)! = n!(n+1)(n+2) > n!n^2, \ ...,$$

then

$$\Delta = r_n < \frac{2}{(2n+3)n!} + \frac{2}{(2n+3)n!n} + \frac{2}{(2n+3)n!n^2} + ... =$$

$$= \frac{2}{(2n+3)n!} \cdot \left(1 + \frac{1}{n} + \frac{1}{n^2} + \ldots\right) = \frac{2}{(2n+3)n!} \cdot \frac{1}{1 - \frac{1}{n}} = \frac{2}{(2n+3)(n-1)!(n-1)}$$

because the expression in parentheses is the sum of a convergent geometric series.

We calculate the error that occurs when $n = 7$:

$$\Delta = r_n < \frac{2}{17 \cdot (1 \cdot 2 \cdot 3 \cdot 4 \cdot 5 \cdot 6) \cdot 6} \approx 0{,}00\,003 < 0{,}0001.$$

Thus, to ensure the desired accuracy of the integral calculation, it is necessary to take the first seven terms:

$$\int_0^1 \sqrt{x}e^x\,dx \approx$$

$$\approx 0{,}66\,667 + 0{,}40\,000 + 0{,}14\,286 + 0{,}03\,704 + 0{,}00\,758 + 0{,}00\,128 +$$

$$+ 0{,}00\,018 = 1{,}25\,561 \approx 1{,}2556.$$

Example 3. Calculate the integral $\displaystyle\int_0^1 \frac{\sin x}{x}\,dx$ with the accuracy up to 10^{-4}.

Solution. It is known that

$$\sin x = x - \frac{x^3}{3!} + \frac{x^5}{5!} - \frac{x^7}{7!} + \ldots.$$

By dividing the term by term by x, we obtain:

$$\frac{\sin x}{x} = 1 - \frac{x^2}{3!} + \frac{x^4}{5!} - \frac{x^6}{7!} + \ldots.$$

By applying the term by term integration, we find:

$$\int_0^1 \frac{\sin x}{x}dx = \int_0^1 \left(1 - \frac{x^2}{3!} + \frac{x^4}{5!} - \frac{x^6}{7!} + \dots\right)dx = \left.\left(x - \frac{x^3}{3!\cdot 3} + \frac{x^5}{5!\cdot 5} - \frac{x^7}{7!\cdot 7} + \dots\right)\right|_0^1$$

Thus, the function $Si\,x$ is defined as the sum of:

$$\int_0^1 \frac{\sin x}{x}dx = 1 - \frac{1}{3!\cdot 3} + \frac{1}{5!\cdot 5} - \frac{1}{7!\cdot 7} + \dots .$$

The fourth term of the series is $\dfrac{1}{7!\cdot 7} \approx 0{,}00\,003 < 10^{-4}$, and the third

term of the series is $\dfrac{1}{5!\cdot 5} \approx 0{,}00166 > 10^{-4}$. Therefore, to calculate the

integral with a given accuracy is sufficient to find the sum of only the first three terms, the value will be calculated "more abundantly."

$$\int_0^1 \frac{\sin x}{x}dx = 1 - \frac{1}{18} + \frac{1}{600} = 0{,}9461.$$

Exercises

1. Express the definite integral in the form of a convergent series, using the Maclaurin series for the integrand. Find the approximate value of this integral with the accuracy up to 10^{-3}.

1.1. $\displaystyle\int_0^{1/2} e^{-x^3} dx$.

1.2. $\displaystyle\int_0^{1/4} \sin\left(x^2\right)dx$.

1.3. $\displaystyle\int_0^{0,1} \frac{\ln(1+x)}{x}dx$.

1.4. $\displaystyle\int_2^4 e^{1/x} dx$.

1.5. $\displaystyle\int_0^1 \frac{\sin x}{\sqrt{x}}dx$.

1.6. $\displaystyle\int_0^{0,2} \frac{\ln(1+5x)}{x}dx$.

1.7. $\displaystyle\int_0^{0,5} xe^{-x} dx$.

1.8. $\displaystyle\int_0^{3/5} \frac{\sin\left(x^2\right)}{x}dx$.

1.9. $\displaystyle\int_5^{10} \frac{\ln(1+x^2)}{x^2}dx$.

1.10. $\int\limits_{1}^{2}\dfrac{e^x}{x}dx$.   **1.11.** $\int\limits_{0}^{1}\cos\sqrt{x}dx$.   **1.12.** $\int\limits_{0}^{0,8}\dfrac{dx}{1+x^5}$.

1.13. $\int\limits_{0}^{4}\sqrt{1+x^3}\,dx$.   **1.14.** $\int\limits_{0}^{0,25}\dfrac{\cos 4x}{\sqrt{x}}dx$.   **1.15.** $\int\limits_{0}^{0,2}\dfrac{dx}{\sqrt{1+x^2}}$.

1.16. $\int\limits_{0}^{0,1}\sqrt[3]{1+3x}dx$.   **1.17.** $\int\limits_{0}^{0,5}\cos\!\left(x^2\right)\!dx$.   **1.18.** $\int\limits_{0}^{0,8}\dfrac{\operatorname{arctg}(1,25x)}{x}dx$.

1.19. $\int\limits_{0}^{0,6}\sqrt[3]{1+x^2}\,dx$.   **1.20.** $\int\limits_{1/4}^{1/2}\dfrac{\cos\dfrac{x^2}{2}}{x^2}dx$.   **1.21.** $\int\limits_{0}^{0,2}\dfrac{\arcsin 5x}{x^2}dx$.

1.22. $\int\limits_{0}^{1}\sqrt[3]{x}\cos xdx$ **1.23.** $\int\limits_{0}^{0,2}\dfrac{\operatorname{arctg} x}{x}dx$ **1.24.** $\int\limits_{0}^{0,5}\dfrac{\arcsin 2x}{\sqrt{x}}dx$

2. Write a computer program to solve one of the exercises of Section 1 above.

3. Application of a power series for finding particular solutions of differential equations

When differential equations cannot be solved by known methods, integration is done with approximation methods. One of such methods is the representation of the solution of an equation in the form of a Taylor series. The sum of a finite number of terms of this series will produce an approximate value of the desired particular solution.

Consider two methods of solving differential equations using power series.

The first - *successive differentiation* - used for solving differential equations of any order.

For example, suppose you want to find a solution to the differential equation of the second order

$$y'' = F(x, y, y'),$$ (22)

satisfying the initial conditions

$$y\big|_{x=x_0} = y_0, \quad y'\big|_{x=x_0} = y_0'.$$ (23)

Let's assume that the solution of $y = f(x)$ exists and can be represented as a Taylor series

$$y = f(x) = f(x_0) + \frac{f'(x_0)}{1!}(x - x_0) + \frac{f''(x_0)}{2!}(x - x_0)^2 + ...,$$ (24)

and the first two terms are known from the initial conditions (23):

$$f(x_0) = y_0,$$

$$f'(x_0) = y_0'.$$

From the equation (22) we find

$$f''(x_0) = y''\big|_{x=x_0} = F(x_0, y_0, y_0').$$

By differentiating both sides of equation (22) by x, one can find any number of derivatives of the unknown function $f(x)$ at x_0. The first differentiation leads to the expression:

$$y''' = F_x'(x, y, y') + F_y'(x, y, y')y' + F_{y'}'(x, y, y')y''.$$ (25)

By substituting the value of $x = x_0$ in the right side of (25), we find

$$f'''(x_0) = y'''\Big|_{x=x_0} \, .$$

Differentiation of (25) will result in

$$f^{IV}(x_0) = y^{IV}\Big|_{x=x_0} \, ,$$

etc.

By placing the obtained derivatives of the function $f(x)$ in the expansion (24), we obtain:

$$y = y_0 + y_0'(x - x_0) + \frac{F(x_0, y_0, y_0')}{2!}(x - x_0)^2 + \dots \, .$$

For the values of x, for which the series converges, it is the desired solution.

Note that the method of successive derivatives in general, does not give the opportunity to explore the convergence of the resulting series to the solution due to the fact that in many cases it is not possible to find an analytical expression for the general term of the sought for series. However, this method is applicable, if it is known that a solution exists in the form of a series. Very often it is used in engineering practice, research, when a decision can be tested experimentally.

Example 1. Find a solution of the equation $y'' = -yx^2$, satisfying the initial conditions $y\big|_{x=0} = 1$, $y'\big|_{x=0} = 0$.
Solution. We have

$$f(0) = y_0 = 1,$$

$$f'(0) = y_0' = 0.$$

From the original equation we find the value $f''(0)$

$$f''(0) = y''\big|_{x=0} = -0 \cdot 0^2 = 0.$$

To find the subsequent coefficients of the series, we differentiate the original equation $y'' = -yx^2$ and at $x = 0$ we get:

$$y''' = -y'x^2 - 2xy,$$

$$f'''(0) = y'''\big|_{x=0} = 0,$$
$$y^{IV} = -y''x^2 - 4xy' - 2y,$$

$$f^{IV}(0) = y^{IV}\big|_{x=0} = -2.$$

By differentiating both parts of the original equation k times by Leibniz's formula, we find

$$y^{(k+2)} = -y^{(k)}x^2 - 2kxy^{(k-1)} - k(k-1)y^{(k-2)}.$$

By assigning $x = 0$, we have:
$$y_0^{(k+2)} = -k(k-1)y_0^{(k-2)}$$
or, by assuming $k + 2 = n$, then

$$y_0^{(n)} = -(n-3)(n-2)y_0^{(n-4)}.$$

Hence,

$$f^{IV}(0) = y_0^{IV} = -1 \cdot 2;$$

$$f^{(8)}(0) = y_0^{(8)} = -5 \cdot 6 y_0^{(4)} = (-1)^2 \cdot (1 \cdot 2) \cdot (5 \cdot 6);$$

$$f^{(12)}(0) = y_0^{(12)} = -9 \cdot 10 y_0^{(8)} = (-1)^3 \cdot (1 \cdot 2) \cdot (5 \cdot 6) \cdot (9 \cdot 10);$$

... ...

$$f^{(4k)}(0) = y_0^{(4k)} = (-1)^k \cdot (1 \cdot 2) \cdot (5 \cdot 6) \cdot (9 \cdot 10) \cdot ... \cdot [(4k-3)(4k-2)].$$

In addition, it can be shown that the derivatives vanish if their order is not multiple of four.

By putting the obtained values of the derivatives in the Maclaurin series, we get:

$$y = 1 - \frac{1 \cdot 2}{4!} x^4 + \frac{(1 \cdot 2) \cdot (5 \cdot 6)}{8!} x^8 - \frac{(1 \cdot 2) \cdot (5 \cdot 6) \cdot (9 \cdot 10)}{12!} x^{12} + \ldots$$

$$\ldots + (-1)^k \frac{(1 \cdot 2) \cdot (5 \cdot 6) \cdot (9 \cdot 10) \cdot \ldots \cdot [(4k-3)(4k-2)]}{(4k)!} x^{4k} + \ldots.$$

With an aid of D'Alembert's criterion, it can be shown that the resulting series converges for all values x and, therefore, is a desired solution of the equation.

Example 2. Find the first six terms of the series expansion of the solution of the equation $y'' = x \sin y'$, satisfying the initial conditions $y\big|_{x=1} = 0$, $y'\big|_{x=1} = \frac{\pi}{2}$.

Solution. The point $x = 1$ is not a particular part of the equation, so its solution can be found in the form of a Taylor series:

$$y = f(1) + \frac{f'(1)}{1!} (x-1) + \frac{f''(1)}{2!} (x-1)^2 + \frac{f'''(1)}{3!} (x-1)^3 + \ldots .$$

Here $f(1) = 0$, $f'(1) = \frac{\pi}{2}$. We calculate 2nd, 3rd, 4th and 5th derivatives and their values at point $x = 1$:

$$f''(1) = 1 \cdot \sin \frac{\pi}{2} = 1,$$

$$f'''(x) = \sin y' + xy'' \cos y' = \sin y' + x \cdot x \sin y' \cdot \cos y' = \sin y' + \frac{x^2 \sin 2y'}{2};$$

$$f'''(1) = 1.$$

Similar to calculations of the previous derivatives, we find the values of the derivatives of the fourth and fifth orders $f^{IV}(1) = -1$, $f^{V}(1) = -6$. By inserting the values of the obtained derivatives in the sought-for series, we obtain the solution of the equation

$$y = \frac{\pi}{2}(x-1) + \frac{1}{2}(x-1)^2 + \frac{1}{6}(x-1)^3 - \frac{1}{24}(x-1)^4 - \frac{1}{20}(x-1)^5 - \dots .$$

Note that it becomes more difficult to find each subsequent derivative of the function $x \sin y'$ than the previous one. In this example, it is very difficult to find the formula of the general term of the sought-for series.

The second method for solving differential equations using power series - *the method of undetermined coefficients* - applies to the equations of any order.

Most often, this method is used to solve linear partial differential equations, and the following theorems are valid for them.

Theorem 1. If in the linear differential equation,

$$p_n(x)y^{(n)} + p_{n-1}(x)y^{(n-1)} + \dots + p_2(x)y'' + p_1(x)y' + p_0(x)y = \varphi(x)$$

the functions $p_k(x)$ ($k = 0, 1, 2, \dots, n$) and $\varphi(x)$ are functions, decomposable in the interval $|x - x_0| < R_1$ in power series by degrees $x - x_0$, and $p_n(x) \neq 0$ in the interval $|x - x_0| < R_2$, than in the interval $|x - x_0| < r$, where $r = \min(R_1, R_2)$, there exists a unique solution $y = f(x)$ of this equation that satisfies the initial conditions $y\big|_{x=x_0} = y_0$, $y'\big|_{x=x_0} = y_0'$, ..., $y^{(n-1)}\big|_{x=x_0} = y_0^{(n-1)}$, where y_0, y_0', ..., $y_0^{(n-1)}$ – arbitrary numbers as a convergent power series:

$$y = a_0 + a_1(x - x_0) + a_2(x - x_0)^2 + \ldots + a_n(x - x_0)^n + \ldots = \sum_{n=0}^{\infty} a_n(x - x_0)^n \ (26)$$

the coefficients a_n of which are yet to be determined.

Theorem 2. If a linear second order differential equation can be represented as

$$y''(x - x_0)^2 + y'(x - x_0)\sum_{n=0}^{\infty} p_n(x - x_0)^n + y\sum_{n=0}^{\infty} q_n(x - x_0)^n = 0,$$

where $\sum_{n=0}^{\infty} p_n(x - x_0)^n$ and $\sum_{n=0}^{\infty} q_n(x - x_0)^n$ – converging in the interval $|x - x_0| < R$ power series in which the coefficients p_0, q_0 and q_1 are not equal to zero at the same time, then there is at least one particular solution $y = f(x)$ of this differential equation in the form of a generalized power series converging in the interval $|x - x_0| < R$:

$$y = \sum_{n=0}^{\infty} a_n(x - x_0)^{n+\rho}, \qquad (27)$$

where ρ is not necessarily an integer number and is determined together with the coefficients a_n of the power series.

The algorithm of the method of undetermined coefficients. In this method, for finding the coefficients of the expansion of a private solution, a power series

$$y = a_0 + a_1 x + a_2 x^2 + \ldots + a_n x^n + \ldots$$

is inserted directly to the original differential equation or the series (26)–(27) and equate the coefficients of the same powers x in different parts of the equation.

Example 3. Find a solution of the equation $y'' = 2xy' + 4y$, satisfying the initial conditions $y\big|_{x=0} = 0$, $y'\big|_{x=0} = 1$.

Solution. We set up

$$y = a_0 + a_1 x + a_2 x^2 + \ldots + a_n x^n + \ldots \; .$$

Based on the initial conditions, we find:

$$a_0 = 0, \; a_1 = 1.$$

Therefore,

$$y = x + a_2 x^2 + a_3 x^3 + \ldots + a_n x^n + \ldots,$$

$$y' = 1 + 2a_2 x + 3a_3 x^2 + \ldots + na_n x^{n-1} + \ldots,$$

$$y'' = 2a_2 + 3 \cdot 2a_3 x + \ldots + n(n-1)a_n x^{n-2} + \ldots \; .$$

By putting the expressions written in the given equation and equating coefficients of the same powers x, we get:

$$2a_2 = 0 \Rightarrow \quad a_2 = 0,$$

$$3 \cdot 2a_3 = 2 + 4 \Rightarrow \quad a_3 = 1,$$

$$4 \cdot 3a_4 = 4a_2 + 4a_2 \Rightarrow \quad a_4 = 0,$$

... ...

$$n(n-1)a_n = (n-2) \cdot 2a_{n-2} + 4a_{n-2} \Rightarrow \quad a_n = \frac{2a_{n-2}}{n-1}.$$

Therefore,

$$a_5 = \frac{2 \cdot 1}{4} = \frac{1}{2}, \ a_6 = \frac{2 \cdot 0}{5} = 0, \ a_7 = \frac{2 \cdot \dfrac{1}{2}}{6} = \frac{1}{3!}, \ a_8 = \frac{2 \cdot 0}{7} = 0,$$

$$a_9 = \frac{2 \cdot \dfrac{1}{3!}}{8} = \frac{1}{4!} \ ...$$

$$... \ a_{2k} = 0, \ a_{2k+1} = \frac{2 \cdot \dfrac{1}{(k-1)!}}{2k} = \frac{1}{k!} \ ... \ .$$

By substituting these coefficients, we obtain the desired solution

$$y = x + \frac{x^3}{1!} + \frac{x^5}{2!} + \frac{x^7}{3!} + ... + \frac{x^{2k+1}}{k!} + ...$$

or, carry $\dfrac{x^3}{1!}$ out of the bracket using the expansion of the function

e^{x^2}, we express the determined partial solution in terms of elementary functions:

$$y = x\left(\frac{x^2}{1!} + \frac{x^4}{2!} + \frac{x^6}{3!} + ... + \frac{x^{2k}}{k!} + ... \right) = xe^{x^2}.$$

Exercises

1. Determine the solutions of the equations by means of consecutive differentiations that meet the specified conditions:

1.1. $y'' = 2xy' + 4y$, $y(0) = 1$, $y'(0) = 1$.

1.2. $y'' = -x^2 y' - 2xy + 1$, $y(0) = 0$, $y'(0) = 0$.

2. Using the method of consecutive differentiations, find a specified number of terms (other than zero) of the expansion of the solutions of the following differential equations at the specified initial conditions.

2.1. $y' = 2\cos x - xy^2$, $y(0) = 1$ (five terms).

2.2. $y'' = xy + y^2$, $y(0) = 1$, $y'(0) = -1$ (five terms).

2.3. $y'' = -2xy$, $y(0) = 1$, $y'(0) = 1$ (five terms).

2.4. $y'' = xyy'$, $y(0) = 1$, $y'(0) = 1$ (six terms).

2.5. $y''' = y'' + (y')^2 + y^3 + x$, $y(0) = 1$, $y'(0) = 2$, $y''(0) = 1/2$ (six terms).

2.6. $y^{IV} = xy + y'x^2$, $y(0) = y'(0) = y''(0) = y'''(0) = 1$ (seven terms).

CHAPTER 4. Numerical solution of algebraic equations $f(x) = 0$

In this chapter, we'll discuss equations with one unknown. Simple linear or quadratic equations can be easily solved with the help of a computer program for calculating formulas. Although algebraic equations of the third and fourth degree can still be solved by analytical methods, the corresponding formulas are so complex that numerical methods have undeniable advantages. For instance, the following equations are cannot be solved in elementary functions:

$$5x^6 - 7x^5 + 3x^4 - 2x^3 + 11x^2 - 4x + 16 = 0;$$

$$x^2 = \cos x.$$

This chapter will cover the following types of equations: *algebraic equations,* for example:

$$x^9 + 2x^6 - 7x^4 - 11x + 33 = 0;$$

transcendental equations, for example

$$15x^3 = e^{x^3};$$

implicit functions, for example:

$$e^{-x^2 - t^2} = x^2 t + t^2 x.$$

Implicit functions are more complex types of equations as include two unknowns. Solution of this type of equation is a table containing all the pairs $x, t,$ that satisfy this equation. Function that cannot be represented as a combination of elementary functions are depicted graphically by plotting the values of x, t in the diagram.

To find a general method for solving the equations of these three types let's present them in a unified or normal form:

$$f(x) = 0.$$

Algebraic equation in its original form is already in this kind.

Transcendental equation in this "normal form" is as follows:

$$f(x) = 15x^3 - e^{x^3} = 0,$$

and the implicit function is:

$$f(x) = e^{-x^2 - t^2} - x^2 t - t^2 x = 0.$$

This expression should be considered as a function of x; t parameter is set to the specified value. Then the expression specified by the implicit function, will be a function of a single variable x. This function on a given interval may be zero, one or several times, or not be zero at all. Thus, as a result of simple transformations, the general problem is reduced to finding such values of the argument for which the function $f(x)$ vanishes.

Numerical methods for the solution, which will be discussed below, apply only to the continuous and differentiable on the interval (where the roots of the equation are sought) functions.

1. Basic definitions and theorems

Let us consider the equation of the form $f(x) = 0$, where $f(x)$ – a function of a real argument, defined and continuous in a finite or infinite interval (a, b).

Definition. The number x_0 from the domain of the function $f(x)$ is called a root of the equation $f(x) = 0$ if $f(x_0) = 0$.

The process of finding the roots of the equation is divided into several stages:

1) There are defined the boundaries of the interval, which contains all the roots of $f(x) = 0$;

2) There are established possibly short intervals, each of which contains exactly one root;

3) Each of the roots is calculated with a predetermined accuracy.

However, in general definition of the boundaries of the interval in which there are all the roots of $f(x) = 0$ one can only determine an algebraic equation in the canonical form, i.e., for the equation:

$$a_0 x^n + a_1 x^{n-1} + a_2 x^{n-2} + \ldots + a_{n-1} x + a_n = 0 \text{, where } a_n \neq 0. \quad (28)$$

In the future, we will only find the real roots of algebraic equations.

Starting from *the first phase of* the process of finding the roots of the equations, we consider a number of theorems without proof that can help determine the number and boundaries of the roots of the algebraic equations (28).

Theorem 1 (the fundamental theorem of algebra). The equation of the type (28) has *n* roots, real or complex, if a root of multiplicity *k* is considered as *k roots.*

Definition. The number x_0 is called a root of multiplicity *k* of the equation (28), if at $x=x_0$ the function $f(x)$ *itself* and its derivatives up to *(k–1)*-th order vanish, i.e.

$$f(x_0) = f'(x_0) = f''(x_0) = \ldots = f^{(k-1)}(x_0) = 0 \text{, and } f^{(k)}(x_0) \neq 0.$$

A root of multiplicity k = 1 is called simple.

.

Theorem 2. 1) The number of real roots of the equation (28) of even degree with real coefficients is always even (also may be zero).

If, in addition, $\dfrac{a_n}{a_0} < 0$, then an equation of even degree has at least two real roots of different sign.

2) The equation (28) of odd degree has at least one real root of the same sign as $-\dfrac{a_n}{a_0}$.

Theorem 3 (Descartes' theorem). The number of positive roots of the equation (28) is equal to, or is an even number smaller than, the number of sign changes in the series of the coefficients $a_0, a_1, a_2, ..., a_{n-1}, a_n$ of the equation.

Remark. Since replacing x by −y the roots of the equation (28) change sign, then with the help of this theorem one can assess the number of negative roots.

Example 1. In the equation of odd degree $3x - 2 = 0$, $\left(x_0 = \dfrac{2}{3}\right)$ the coefficients $a_0 = 3$, $a_n = -2$ and $-\dfrac{a_n}{a_0} = -\dfrac{-2}{3} = \dfrac{2}{3} > 0$.

In addition, the number of sign changes is 1.

Therefore, by Theorems 2 and 3, the examined equation has one real positive root.

Example 2. In the equation of odd degree $x^3 - 2x^2 - x + 2 = 0$ (the solutions or roots of this equation are as follows: $x_1 = -1$, $x_2 = 1$, $x_3 = 2$), the coefficients $a_0 = 1$, $a_n = 2$ and $-\dfrac{a_n}{a_0} = -\dfrac{2}{1} = -2 < 0$.

Consequently, by Theorem 2, this equation has at least one real negative root.

The number of sign changes in the equation is two, therefore, by Theorem 3, it has either two or 0 positive real roots.

We estimate the number of real negative roots. To do this, we replace x by y. We obtain the equation $(-y)^3 - 2(-y)^2 - (-y) + 2 = 0$, or $-y^3 - 2y^2 + y + 2 = 0$, or $y^3 + 2y^2 - y - 2 = 0$. The number of sign changes in this equation is equal to 1, therefore, the original equation has one real negative root.

Example 3. In the equation of an even degree $x^2 - 2x - 3 = 0$ ($x_1 = -1$, $x_2 = 3$) the coefficients $a_0 = 1$, $a_n = -3$ and $\dfrac{a_n}{a_0} = \dfrac{-3}{1} = -3 < 0$.

Therefore, by Theorem 2 the above equation has two real roots of opposite sign.

Example 4. In the equation of an even degree $x^4 + 3x^2 - 4 = 0$, ($x_1 = -1$, $x_2 = 1$), the coefficients $a_0 = 1$, $a_n = -4$ and $\dfrac{a_n}{a_0} = \dfrac{-4}{1} = -4 < 0$. Therefore, by Theorem 2, the equation has at least two real roots of opposite sign.

The number of sign changes in this equation is 1, so by Theorem 3, it has one positive real root.

We estimate the number of real negative roots. To do this, we replace x by y. We obtain the equation $(-y)^4 + 3(-y)^2 - 4 = 0$ or $y^4 + 3y^2 - 4 = 0$. The number of sign changes in the resulting equation is 1, therefore, the original equation has one real negative root.

Let's formulate a theorem that allows, approximately enough, for determining the boundaries of the interval which contains all the real roots of the equation (28).

Theorem 4. *1)* If

$$A = \max|a_i|, \text{ where } 0 \le i \le n-1,$$

$$B = \max|a_i|, \text{ where } 1 \le i \le n,$$

and

$$r = \frac{|a_n|}{A + |a_n|}; \quad R = 1 + \frac{B}{|a_0|},$$

then

$$r \le |x| \le R.$$

2) All of the positive real roots of the equation (28) are in the interval $r \le x \le R$, and all of the negative real roots of the equation (28) are in the interval $-R \le x \le -r$.

Example 5. In the equation $x^2 - 2x - 3 = 0$ $(x_1 = -1, \ x_2 = 3)$ the coefficients are $a_0 = 1$, $a_1 = -2$, $a_2 = a_n = -3$.

We will find the numbers A, B, r, R:

$$A = \max(|a_0|, |a_1|) = \max(1; 2) = 2;$$
$$B = \max(|a_1|, |a_2|) = \max(2; 3) = 3;$$

$$r = \frac{3}{2+3} = 3/5; \quad R = 1 + \frac{3}{1} = 4.$$

All of the positive real roots of the equation are in the interval $3/5 \le x \le 4$ (the value of $x_2 = 3$ falls within this interval).

All of the negative real roots of the equation are in the interval $-4 \le x \le -3/5$ (the value of $x_1 = -1$ falls within this interval).

Example 6. In the equation $x^3 - 2x^2 - x + 2 = 0$, ($x_1 = -1$, $x_2 = 1$, $x_3 = 2$), the coefficients are $a_0 = 1$, $a_1 = -2$, $a_2 = -1$, $a_3 = a_n = 2$.

We will find the numbers A, B, r, R:

$$A = \max\big(|a_0|, |a_1|, |a_2|\big) = \max(1; 2; 1) = 2;$$
$$B = \max\big(|a_1|, |a_2|, |a_3|\big) = \max(2; 1; 2) = 2;$$

$$r = \frac{2}{2+2} = 1/2; \quad R = 1 + \frac{2}{1} = 3.$$

All of the positive real roots of the equation are in the interval $1/2 \le x \le 3$ (the values of $x_2 = 1$ and $x_3 = 2$ fall into this range).

All of the negative real roots of the equation are in the interval $-3 \le x \le -1/2$ (the value of $x_1 = -1$ falls within this interval).

Example 7. In the equation $x^4 + 3x^2 - 4 = 0$, ($x_1 = -1$, $x_2 = 1$), the coefficients are $a_0 = 1$, $a_1 = 0$, $a_2 = 3$, $a_3 = 0$, $a_4 = a_n = -4$.

We will find the numbers A, B, r, R:

$$A = \max\big(|a_0|, |a_1|, |a_2|, |a_3|\big) = \max(1; 0; 3; 0) = 3;$$
$$B = \max\big(|a_1|, |a_2|, |a_3|, |a_4|\big) = \max(0; 3; 0; 4) = 4;$$

$$r = \frac{4}{3+4} = 4/7; \quad R = 1 + \frac{4}{1} = 5.$$

All of the positive real roots of the equation are in the interval $4/7 \le x \le 5$ (the value of $x_2 = 1$ falls within this interval).

All of the negative real roots are in the interval $-3 \le x \le -1/2$ (the value of $x_1 = -1$ falls within this interval).

So, we have learned to find the number of real roots of an algebraic equation, and approximate enough intervals in which they are contained.

The second stage of the process of finding the roots of an algebraic equation sets probably small intervals containing exactly one real root of the algebraic equation.

Now, we formulate the following theorem.

Theorem 5. If a continuous and differentiable function $f(x)$, determining the algebraic equation $f(x) = 0$, takes values of different signs, that is, $f(a)f(b) < 0$ at the ends of the segment $[a; b]$, and its first derivative retains its sign within this interval, then there is exactly one real root of this equation within $[a; b]$.

Example 8. For the equation $x^2 - 2x - 3 = 0$ we will find the intervals of constant sign of a continuous function $f(x) = x^2 - 2x - 3$. For this, we compute the derivative: $f'(x) = 2x - 2$; $f'(x) = 0$ at $2x - 2 = 0$, i.e. $x = 1$.

For $x < 1$ $f'(x) < 0$, i.e. , the function is decreasing.

For $x > 1$ $f'(x) > 0$, i.e. , the function is increasing.

Let's compare these intervals with the previously found intervals, which contain all of the positive and negative real roots of the equation (Fig.6).

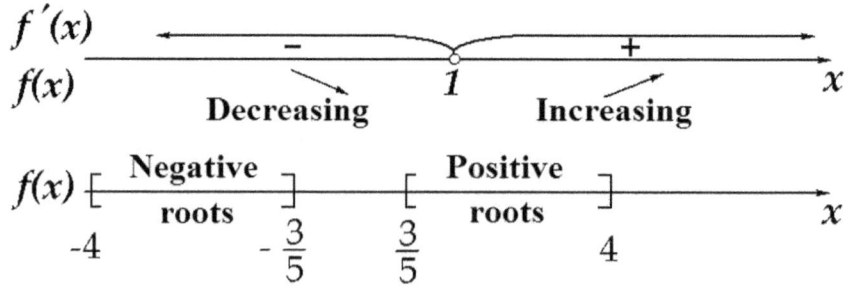

Fig. 6

A comparison of the intervals gets three intervals: $-4 \le x \le -3/5$, $3/5 \le x < 1$ и $1 < x \le 4$.

We find signs of the function at the ends of these intervals:

$$f(-4) = (-4)^2 - 2 \cdot (-4) - 3 = 21 > 0;$$

$$f(-3/5) = (-3/5)^2 - 2 \cdot (-3/5) - 3 = -36/25 < 0,$$

therefore, by Theorem 5 the equation $x^2 - 2x - 3 = 0$, in the interval $-4 \le x \le -3/5$, has exactly one negative real root;

$$f(3/5) = (3/5)^2 - 2 \cdot (3/5) - 3 = -96/25 < 0;$$

$$f(1) = (1)^2 - 2 \cdot 1 - 3 = -4 < 0,$$

therefore, by Theorem 5 this equation, in the interval $3/5 \le x < 1$, has no real roots;

$$f(1) = (1)^2 - 2 \cdot 1 - 3 = -4 < 0;$$

$$f(4) = (4)^2 - 2 \cdot 4 - 3 = 5 > 0,$$

therefore, by Theorem 5 the equation $x^2 - 2x - 3 = 0$, in the interval $1 < x \le 4$, has exactly one positive real root.

Example 9. For the equation $x^3 - 2x^2 - x + 2 = 0$, we find intervals of constant sign of the continuous function $f(x) = x^3 - 2x^2 - x + 2$. For this, we compute the derivative: $f'(x) = 3x^2 - 4x - 1$; $f'(x) = 0$ at $3x^2 - 4x - 1 = 0$, i.e.

$$x_1 = \frac{2 - \sqrt{7}}{3} \approx -0{,}215; \; x_2 = \frac{2 + \sqrt{7}}{3} \approx -1{,}549.$$

At $x \in \left(-\infty; \dfrac{2-\sqrt{7}}{3} \right) \cup \left(\dfrac{2+\sqrt{7}}{3}; +\infty \right)$ the derivative is $f'(x) > 0$, i.e.

the function is increasing; at $x \in \left(\dfrac{2-\sqrt{7}}{3}; \dfrac{2+\sqrt{7}}{3} \right)$ $f'(x) < 0$, i.e.

the function is decreasing (Fig. 7).

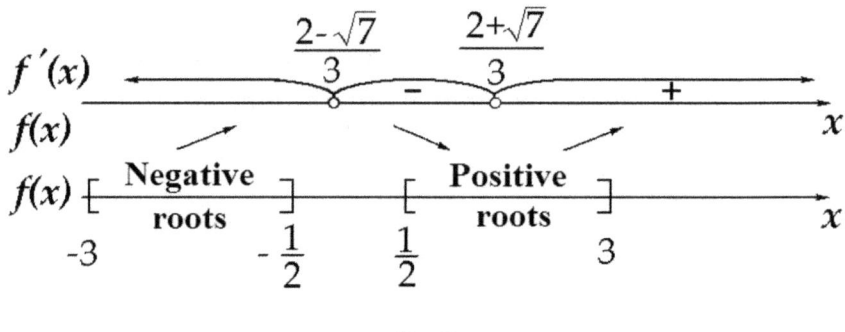

Fig. 7

We compare these intervals with the previously found intervals which contain all the positive and negative real roots of the equation (Fig.7). From this comparison, we get three intervals:

$$-3 \le x \le -1/2, \; 1/2 \le x < \dfrac{2+\sqrt{7}}{3} \text{ and } \dfrac{2+\sqrt{7}}{3} < x \le 3.$$

We find signs of the function at the ends of these intervals:

$$f(-3) = (-3)^3 - 2 \cdot (-3)^2 - (-3) + 2 = -40 < 0;$$

$$f(-1/2) = (-1/2)^3 - 2 \cdot (-1/2)^2 - \left(-\dfrac{1}{2} \right) + 2 > 0,$$

therefore, by Theorem 5, the equation $x^3 - 2x^2 - x + 2 = 0$, in the interval $-3 \le x \le -1/2$, has exactly one negative real root;

$$f(1/2) = (1/2)^3 - 2 \cdot (1/2)^2 - \left(\frac{1}{2}\right) + 2 > 0;$$

$$f\left(\frac{2+\sqrt{7}}{3}\right) = \left(\frac{2+\sqrt{7}}{3}\right)^3 - 2\left(\frac{2+\sqrt{7}}{3}\right)^2 - \left(\frac{2+\sqrt{7}}{3}\right) + 2 < 0,$$

therefore, by Theorem 5, the equation, in the interval $1/2 \le x < \dfrac{2+\sqrt{7}}{3}$, has exactly one positive real root;

$$f\left(\frac{2+\sqrt{7}}{3}\right) = \left(\frac{2+\sqrt{7}}{3}\right)^3 - 2 \cdot \left(\frac{2+\sqrt{7}}{3}\right)^2 - \left(\frac{2+\sqrt{7}}{3}\right) + 2 < 0;$$

$$f(3) = 3^3 - 2 \cdot 3^2 - 3 + 2 > 0,$$

therefore, by Theorem 5, the equation $x^3 - 2x^2 - x + 2 = 0$, in the interval $\dfrac{2+\sqrt{7}}{3} < x \le 3$, has exactly one positive real root.

Example 10. For the equation $x^4 + 3x^2 - 4 = 0$, we find intervals of constant sign of the continuous function $f(x) = x^4 + 3x^2 - 4$. For this, we compute the derivative: $f'(x) = 4x^3 + 6x$; $f'(x) = 0$ at $4x^3 + 6x = 0$ or $2x(2x^2 + 3) = 0$, therefore $x = 0$.

At $x < 0$ $f'(x) < 0$, i.e. the function is decreasing; at $x > 0$ $f'(x) > 0$, i.e. the function is increasing (Fig. 8).

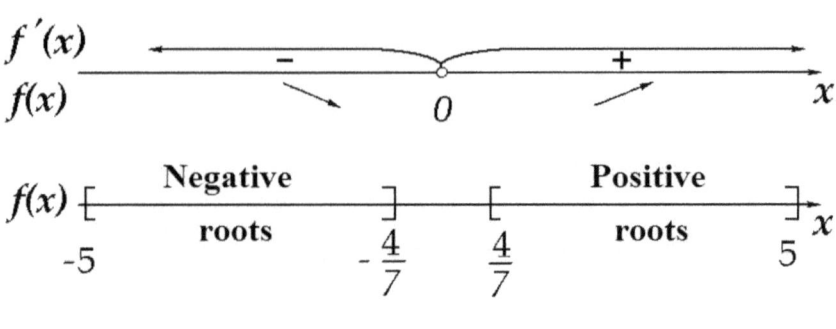

Fig. 8

We compare these intervals with the previously found intervals which contain all the positive and negative real roots of the equation (Fig.8). From their comparison, we obtain two intervals: $-5 \le x \le -4/7$ and $\dfrac{4}{7} \le x \le 5$.

We find signs of the function at the ends of these intervals:

$$f(-5) = (-5)^4 + 3 \cdot (-5)^2 - 4 > 0;$$

$$f(-4/7) = (-4/7)^4 + 3 \cdot (-4/7)^2 - 4 < 0,$$

therefore, by Theorem 5, the equation, in the interval $-5 \le x \le -4/7$, has exactly one negative real root;

$$f(4/7) = (4/7)^4 + 3 \cdot (4/7)^2 - 4 < 0;$$

$$f(5) = 5^4 + 3 \cdot 5^2 - 4 > 0,$$

therefore, by Theorem 5, the equation $x^4 + 3x^2 - 4 = 0$, in the interval $\dfrac{4}{7} \le x \le 5$, has exactly one positive real root.

R e m a r k. For algebraic equations (28), the degrees of which are greater than three, it is difficult to analytically find the intervals of constant sign of the function $f(x)$. Therefore, to find the smallest possible intervals containing exactly one real root, in practice, the following ways are applied:

1) A function $f(x)$ is presented in a graphical form and short intervals are approximately determined that contain exactly one root (i.e. the intervals that contain a single point of intersection of the graph of $f(x)$ with the x-axis);

2) If the graph of the function $f(x)$ is difficult to build, then we form simple functions $\varphi_1(x)$ and $\varphi_2(x)$ such as the equation $f(x) = 0$ is converted into the form of $\varphi_1(x) = \varphi_2(x)$. Then we plot graphs of the functions $y = \varphi_1(x)$ and $y = \varphi_2(x)$ and approximately determine the intervals, containing the abscissas of the points of intersection of these graphs.

For example, the equation $x^3 - 2x^2 - x + 2 = 0$ can be transformed to $x^3 = 2x^2 + x - 2$ a n d then one can build graphs of the functions $y = x^3$ and $y = 2x^2 + x - 2$.

Since *the third stage* of the process of finding the roots of an algebraic equation, we give the statement of the theorem which lets estimate the error of the approximate solutions.

Theorem 6. If x_1 is accurate, and x_2 is approximate roots of (28) belonging to the same interval $[a; b]$, then the estimate $|x_2 - x_1| \leq \dfrac{|f(x_2)|}{m}$ is valid, where m is the minimum value of the modulus of the derivate of the function $f(x)$ in the interval $[a; b]$.

Example 11. Previously, we found that the equation $x^3 - 2x^2 - x + 2 = 0$ in the interval $-3 \leq x \leq -1/2$ has exactly one negative real root.

Let's suppose that its approximate value $x_2 = -1,1$. We estimate the absolute error $|x_2 - x_1|$, where x_1 is the unknown exact value of the root.

For this, we compute $|f(x_2)|$ and m:

a) $f(x_2) = f(-1,1) = (-1,1)^3 - 2 \cdot (-1,1)^2 - (-1,1) + 2 =$

$$= -1,331 - 2,42 + 1,1 + 2 = -0,651;$$

$$|f(x_2)| = |-0,651| = 0,651;$$

b) in the interval $-3 \leq x \leq -1/2$, the derivative $f'(x) = 3x^2 - 4x - 1$, staying positive changes

from the value $f'(-3) = 3 \cdot (-3)^2 - 4 \cdot (-3) - 1 = 38$

to $f'\left(-\dfrac{1}{2}\right) = 3 \cdot \left(-\dfrac{1}{2}\right)^2 - 4 \cdot \left(-\dfrac{1}{2}\right) - 1 = \dfrac{7}{4}$.

Therefore, the minimum value of its modulus is $\dfrac{7}{4}$, i.e. $m = \dfrac{7}{4}$.

According to Theorem 6, $|x_2 - x_1| = |-1,1 - x_1| \leq \dfrac{0,651}{7/4} \approx 0,37$.

Next, based on specific examples, there will be presented several ways to find approximate values of the real roots of the equation (28).

2. The chord method

Let's suppose that we were able to find a small enough interval $[x_1; x_2]$, that has exactly one real root of the equation (28).

According to Theorem 5, a continuous and differentiable function $y = f(x)$ takes different signs on the ends of the interval, i.e.. $f(x_1)f(x_2) < 0$.

Assume also that the interval $[x_1; x_2]$ is so small that at all points $f'(x)$ and $f''(x)$ retain a constant sign.

In Fig. 9-12, we depict four types of the graphics for the location of the arc of the curve.

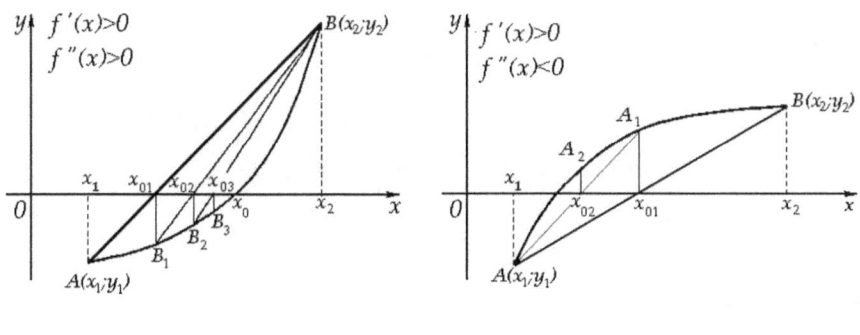

Fig. 9   Fig. 10

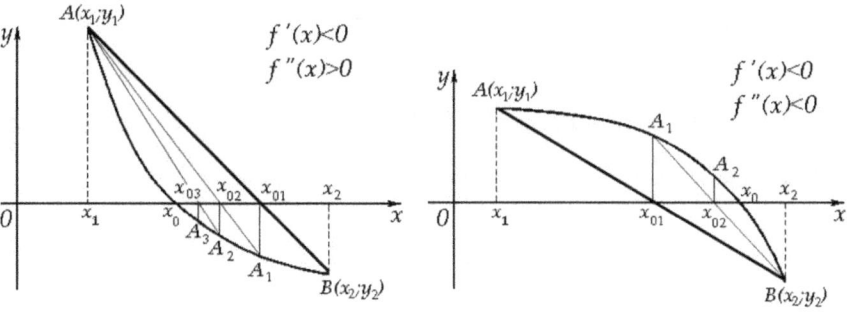

Fig. 11   Fig. 12

We consider and describe separately two cases.

Case 1. $f'(x)f''(x) > 0$ in $[x_1; x_2]$ (Fig. 9 and 12), i.e. either $f'(x) > 0$ and $f''(x) > 0$ in $[x_1; x_2]$ (Fig. 9), either $f'(x) < 0$ and $f''(x) < 0$ in $[x_1; x_2]$ (Fig. 12);

Case 2. $f'(x)f''(x) < 0$ in $[x_1; x_2]$ (Fig. 10 и 11), i.e.. either $f'(x) > 0$ and $f''(x) < 0$ in $[x_1; x_2]$ (Fig. 10), either $f'(x) < 0$ and $f''(x) > 0$ in $[x_1; x_2]$ (Fig. 11).

We present *an algorithm* solving the problem *in the first case:*

a) through the points $A(x_1; y_1)$ and $B(x_2; y_2)$ of the graph $y = f(x)$, we draw a chord *AB*. Its equation is

$$\frac{x - x_1}{x_2 - x_1} = \frac{y - y_1}{y_2 - y_1} \quad \text{or} \quad x = x_1 + \frac{x_2 - x_1}{y_2 - y_1}(y - y_1);$$

b) we find the abscissa of the point of intersection of the chord *AB and* the *x-axis.*

By assigning y = 0, we obtain $x_{01} = x_1 - \dfrac{x_2 - x_1}{y_2 - y_1} y_1$;

c) we put x_{01} into the function $y = f(x)$ and obtain $y_{01} = f(x_{01})$. The point B_1 has the coordinates $B_1(x_{01}; y_{01})$;

d) through the points $B_1(x_{01}; y_{01})$ and $B(x_2; y_2)$ of the graph $y = f(x)$, we draw a chord $B_1 B$. Its equation is

$$\frac{x - x_{01}}{x_2 - x_{01}} = \frac{y - y_{01}}{y_2 - y_{01}} \quad \text{or} \quad x = x_{01} + \frac{x_2 - x_{01}}{y_2 - y_{01}}(y - y_{01});$$

e) we find the abscissa of the point of intersection of the chord B_1B with the x-axis. By assigning $y = 0$, we get:

$$x_{02} = x_{01} - \frac{x_2 - x_{01}}{y_2 - y_{01}} y_{01};$$

f) as a result of the iterations a) – e) for the consequentially calculated values x_{0i} $(i = 2, 3, 4, ...)$, we obtain the sequence of values $x_{01}, x_{02},\ x_{03},\ ...,$ converging to x_0.

After the inequality $|x_{0i+1} - x_{0i}| < \varepsilon$, where ε is a selected by us the accuracy of our approach, is satisfied, the process should be complete.

Thus, in the first case, the calculations are done by the formulas:

$$x_{01} = x_1 - \frac{x_2 - x_1}{y_2 - y_1} y_1; \qquad y_{0i} = f(x_{0i})$$

$$x_{0i+1} = x_{0i} - \frac{x_2 - x_{0i}}{y_2 - y_{0i}} y_{0i}, \quad \text{where} \quad i = 1, 2, 3,... \tag{29}$$

We present *an algorithm* solving the problem in *the second case:*

a)   the values of x_{01} и y_{01} are determined similarly to the first case. The point A_1 has the coordinates $A_1(x_{01}; y_{01})$;

b) through the points A and A_1 of the graph $y = f(x)$, we draw a chord AA_1. Its equation is:

$$\frac{x - x_1}{x_{01} - x_1} = \frac{y - y_1}{y_{01} - y_1} \quad \text{or } x = x_1 + \frac{x_{01} - x_1}{y_{01} - y_1}(y - y_1);$$

c) we find the abscissa of the point of intersection of the chord AA_1

63

with the *x-axis*. By assigning $y = 0$, we will get: $x_{02} = x_1 - \dfrac{x_{01} - x_1}{y_{01} - y_1} y_1$;

d) the further steps are the same as in the first case..

So, in the second case, the calculations are done by the formulas:

$$x_{01} = x_1 - \frac{x_2 - x_1}{y_2 - y_1} y_1; \qquad y_{0i} = f(x_{0i})$$

$$x_{0i+1} = x_1 - \frac{x_{0i} - x_1}{y_{0i} - y_1} y_1; \quad \text{where} \quad i = 1, 2, 3,... \tag{30}$$

Example. By the method of chords, find the approximate value of the positive real root of the equation $x^3 - 2x - 1 = 0$ with the accuracy 0,1.

Solution. 1) It can be shown that the approximate boundaries of an interval containing one positive real root of this equation are $1 \le x_0 \le 2$.

2) $f'(x) = 3x^2 - 2$; $f''(x) = 6x$. As $f'(x)$ and $f''(x) > 0$ at $x \in [1; 2]$, then we perform calculations using the formulas (29). We make a schematic drawing (Fig. 13).

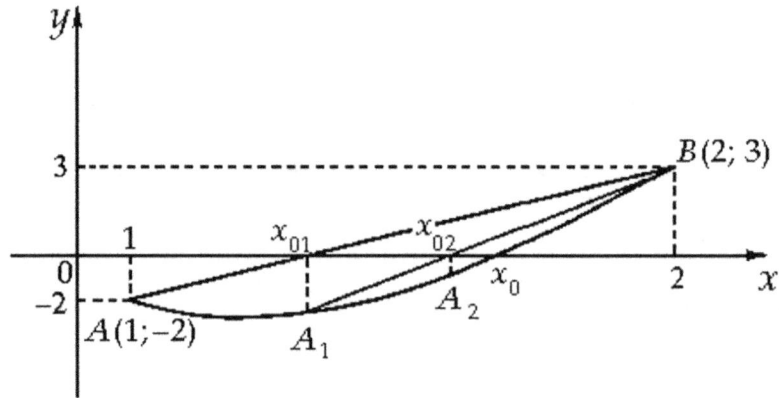

Fig. 13

3) The coordinates of the points A and B: $A(1; -2)$, $B(2; 3)$.

64

Thus, $A(x_1; y_1) = (1; -2)$; $B(x_2; y_2) = (2; 3)$, therefore,

$x_1 = 1$, $x_2 = 2$, $y_1 = -2$, $y_2 = 3$.

4) On the basis of (29) we have:

$$x_{01} = 1 - \frac{2-1}{3+2} \cdot (-2) = 1,4;$$

$y_{01} = f(x_{01}) = (1,4)^3 - 2 \cdot 1,4 - 1 = -1,056$;

$$x_{02} = x_{01} - \frac{x_2 - x_{01}}{y_2 - y_{01}} y_{01} = 1,4 - \frac{2-1,4}{3+1,056} \cdot (-1,056) \approx 1,5562;$$

$$y_{02} = f(x_{02}) = (1,5562)^3 - 2 \cdot 1,5562 - 1 \approx -0,3437.$$

5) We form the difference $|x_{02} - x_{01}| = |1,5562 - 1,4| = 0,1562$. As $0,1562 > 0,1$, then we continue the process of approximation.

6) From (29) we have:

$$x_{03} = x_{02} - \frac{x_2 - x_{02}}{y_2 - y_{02}} y_{02} = 1,5562 - \frac{2-1,5562}{3+0,3437} \cdot (-0,3437) \approx 1,6018.$$

7) We form the difference $|x_{02} - x_{01}| = |1,6018 - 1,5562| = 0,0456$. As $0,0456 < 0,1$, the process of approximation is complete.

Thus, with the accuracy up to 0.1, the approximate value of the positive real root of the equation $x^3 - 2x - 1 = 0$ is 1,6018.

3. Newton's method

Under the same assumptions as in the method of chords in Fig. 14-17, we depict graphs of four types of the location of the arc of the curve.

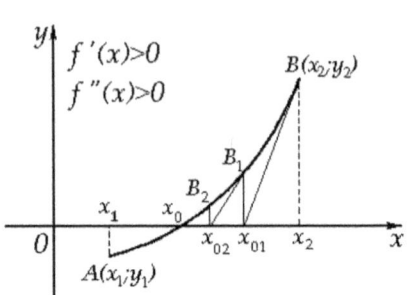

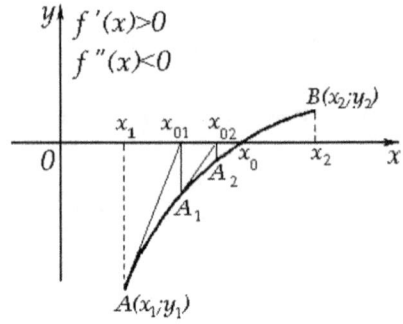

Fig. 14   Fig. 15

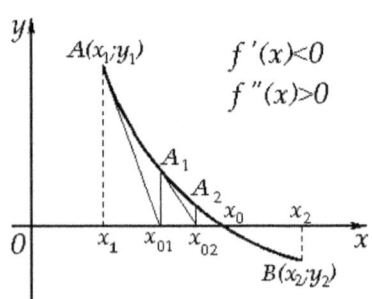

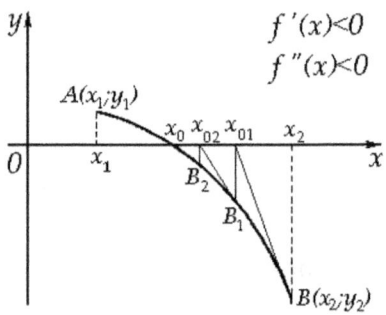

Fig. 16   Fig. 17

We consider and describe two cases separately.

Case 1. $f'(x)f''(x) > 0$ in $[x_1; x_2]$ (Fig. 14 и 17), i.e. either $f'(x) > 0$ and $f''(x) > 0$ in $[x_1; x_2]$ (Fig. 14), either $f'(x) < 0$ and $f''(x) < 0$ in $[x_1; x_2]$ (Fig. 17);

Case 2. $f'(x)f''(x) < 0$ in $[x_1; x_2]$ (Fig. 15 and 16), i.e. either $f'(x) > 0$ and $f''(x) < 0$ in $[x_1; x_2]$ (Fig. 15), either $f'(x) < 0$ and $f''(x) > 0$ in $[x_1; x_2]$ (Fig. 16).

We present *an algorithm* solving the problem *in the first case:*

a) through the point $B(x_2; y_2)$, we draw a tangent to the curve $y = f(x)$. Its equation is:

$$y - y_2 = f'(x_2)(x - x_2) \text{ or } x = x_2 + \frac{1}{f'(x_2)}(y - y_2);$$

b) we find the abscissa of the point of intersection of the tangent with the x-axis. By assigning $y = 0$, we obtain $x_{01} = x_2 - \frac{y_2}{f'(x_2)}$;

c) we put the value of x_{01} in the equation of the curve $y = f(x)$, and we get: $y_{01} = f(x_{01})$. The point B_1 has the coordinates $B_1(x_{01}; y_{01})$;

d) through the point B_1 we draw a tangent to the curve $y = f(x)$. Its equation is:

$$y - y_{01} = f'(x_{01})(x - x_{01}) \text{ or } x = x_{01} + \frac{1}{f'(x_{01})}(y - y_{01});$$

e) we find the abscissa of the point of intersection of the tangent with the x-axis. By assigning $y = 0$, we will get $x_{02} = x_{01} - \frac{y_{01}}{f'(x_{01})}$;

f) as a result we obtain a sequence of the values $x_{01}, x_{02}, x_{03}, \ldots,$ converging to x_0.

After the inequality $|x_{0i+1} - x_{0i}| < \varepsilon$, where ε is selected by us the accuracy of the approach, is satisfied, , the process should be complete.

Thus, in the first case, the calculations are done by the formulas

$$x_{01} = x_2 - \frac{y_2}{f'(x_2)}; \qquad y_{0i} = f(x_{0i});$$

$$x_{0i+1} = x_{0i} - \frac{y_{0i}}{f'(x_{0i})}, \qquad \text{where} \quad i = 1, 2, 3, \dots. \tag{31}$$

An algorithm for solving the problem *in the second case* is the same as in the first case, only the first tangent will be drawn through the point $A(x_1; y_1)$.

So, in the second case, the calculations are done by the formulas:

$$x_{01} = x_1 - \frac{y_1}{f'(x_1)}; \qquad y_{0i} = f(x_{0i});$$

$$x_{0i+1} = x_{0i} - \frac{y_{0i}}{f'(x_{0i})}, \qquad \text{where} \quad i = 1,2,3,\dots. \tag{32}$$

Example. By Newton's method, find the approximate value of a positive real root of the equation $x^3 - 2x - 1 = 0$ with the accuracy 0,1.

Solution. 1) The boundaries of the interval containing one positive real root of this equation are $1 \le x_0 \le 2$.

2) $f'(x) = 3x^2 - 2$; $f''(x) = 6x$. As $f'(x)$ and $f''(x) > 0$ at $x \in [1; 2]$, then we perform calculations using formulas (31). We make a schematic drawing (Fig. 18).

3) The coordinates of the points are $A(x_1; y_1) = (1; -2)$; $B(x_2; y_2) = (2; 3)$, therefore, $x_1 = 1$, $x_2 = 2$, $y_1 = -2$, $y_2 = 3$.

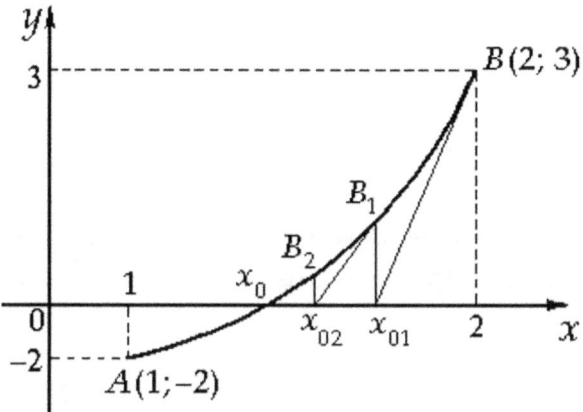

Fig. 18

4) On the basis of (31) we have:

$$f'(2) = 3 \cdot 2^2 - 2 = 10 \, ;$$

$$x_{01} = x_1 - \frac{y_1}{f'(x_1)} = 2 - \frac{3}{10} = 1{,}7 \, ;$$

$$y_{01} = f(x_{01}) = (1{,}7)^3 - 2 \cdot 1{,}7 - 1 = 0{,}513 \, ;$$

$$x_{02} = x_{01} - \frac{y_{01}}{f'(x_{01})} \, ; \quad f'(x_{01}) = f'(1{,}7) = 3 \cdot (1{,}7)^2 - 2 = 6{,}67 \, ;$$

$$x_{02} = 1{,}7 - \frac{0{,}513}{6{,}67} \approx 1{,}623 \, ;$$

$$y_{02} = f(x_{02}) = (1{,}623)^3 - 2 \cdot 1{,}623 - 1 \approx 0{,}029 \, .$$

5) We form the difference $|x_{02} - x_{01}| = |1{,}623 - 1{,}7| = 0{,}077$. As $0{,}077 < 0{,}1$, the process of approximation is complete.

Thus, with the accuracy up to 0.1, the approximate value of the positive real root of the equation $x^3 - 2x - 1 = 0$ is 1,623.

Exercises

1. By chord's method, and with the accuracy 0.01, calculate the approximate value of the largest real root of the following algebraic equations:

1.1. $x^4 - 4x - 1 = 0$. **1.2.** $x^3 - 3x^2 + 5x - 4 = 0$.

1.3. $x^3 - x^2 - 2x + 1 = 0$. **1.4.** $x^3 + 2x^2 - 3x - 7 = 0$.

1.5. $x^3 + 3x^2 + 2x + 5 = 0$. **1.6.** $x^3 - 2x^2 + 3x - 5 = 0$.

2. By Newton's method, and with the accuracy 0.01, calculate the approximate value of the largest real root of the following algebraic equations:

2.1. $2x^3 - 5x^2 + 1 = 0$. **2.2.** $x^3 - 2x^2 - 4x - 7 = 0$.

2.3. $x^3 - x^2 - 9x + 9 = 0$. **2.4.** $5x^3 - x - 1 = 0$.

2.5. $x^3 + x^2 - 3 = 0$. **2.6.** $3x^3 - 0,9x - 6 = 0$.

ANNEX

Euler's method (the trapezoidal rule)
for approximate calculation of integrals

A simplified block diagram of the following program for the calculation of an integral by the method of trapezoids, in its nucleus, contains a loop FOR-NEXT, in which an integral sum is calculated for N intervals. In the outer endless cycle, the number of intervals is doubled after each pass of the inner loop and a current value of the sum (an approximate value of the integral) is displayed.

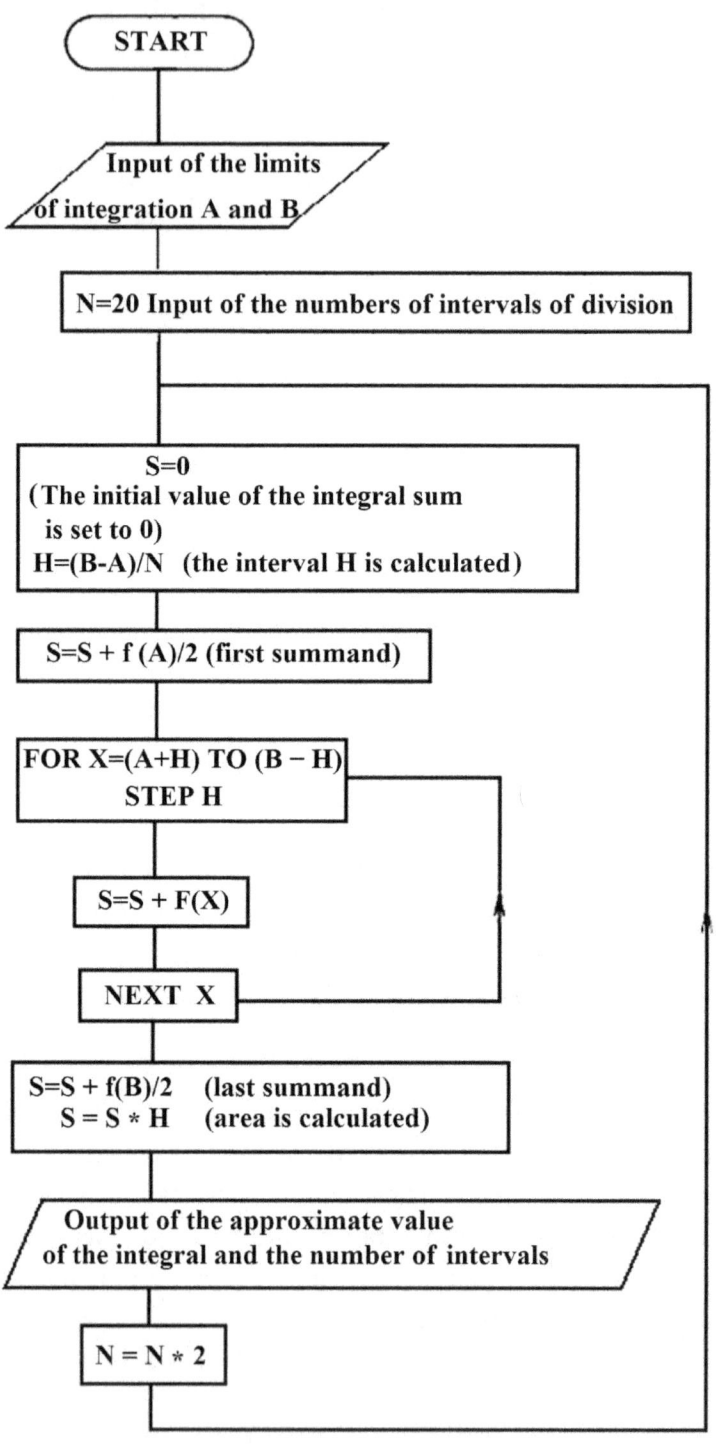

By comparing the calculated and the on-screen approximate values of the integral, you can decide whether you are satisfied with the result. Accordingly, you can manually abort the program.

However, if the program includes calculation of a large number of integrals by iterative (approximate) methods, the interruption of the calculation of each integral manually is not only appropriate but impossible. Therefore, the program needs an exit out of the endless cycles of iteration when it reaches the required accuracy of the results.

Task. Organize a way out of an infinite loop, using the following method. After each iteration, you need to compare the obtained value of the sum with its value obtained in the previous iteration. If the difference between these values is less than a predetermined value, the cyclic procedure is interrupted.

Numerical integration of differential equations of the first order by Euler's method

The block diagram of the algorithm for solving differential equations by Euler's method is shown below (assuming that the program does not provide for automatic exit from the endless cycle).

The block diagram consists of three parts. The first part corresponds to the main program. In it, a subroutine (the second portion of the block diagram) is called out, which implements the algorithm for Euler's method. This subroutine in turn, at each invocation, calls the subroutine "DE" (the third part of the flowchart), which calculates the right-hand side of the differential equation.

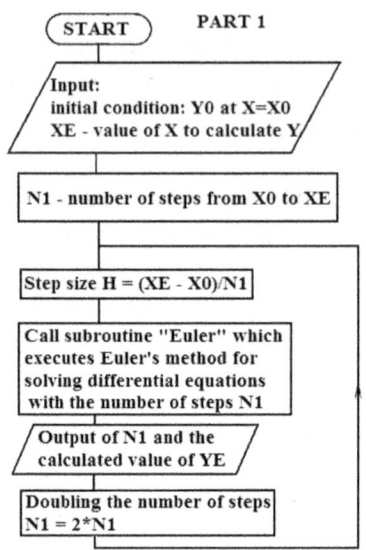

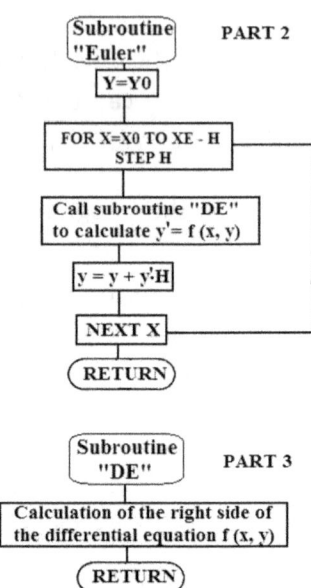

The chord method

The advantage of the present method in comparison with the method of tangents is that the program does not need to enter an analytical expression for the derivative. Disadvantages of the method include the need to set two initial values of the independent variable and, in addition, a slow convergence results.

The flowchart of the algorithm for solving equations by the chord method is presented below.

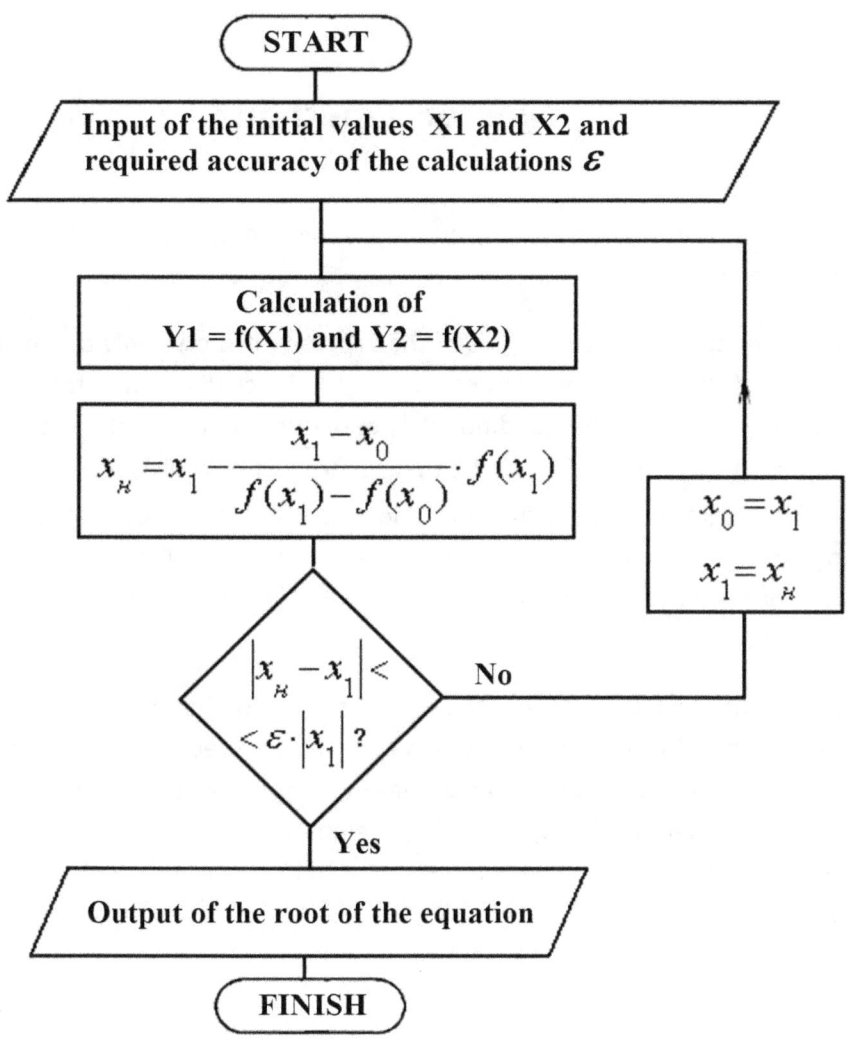

Newton's method

The block diagram for solving equations by the Newton method is shown below.

It does not contain all the details of the program, it shows only the main features of the algorithm. For example, in the block diagram before the iterative procedure, the conditional operator is omitted that tests the condition $f'(x) = 0$. If $f'(x) = 0$, then, there will be a division by zero, and the program will be terminated. Geometrically, this means that the tangent is parallel to the *x-axis,* and does not cross it. The conditional operator operates similarly at finding out whether the desired accuracy is achieved.

The accuracy of the result and the time required to obtain it depend on what numerical value is given to the initial value x. Because of that, the flowchart includes checking the condition if the derivative takes a too small value. Geometrically, this is equivalent to the question whether the slope of the tangent line is too small. If the value of the derivative is too small, then a second request for a more appropriate initial approximation will be prompted.

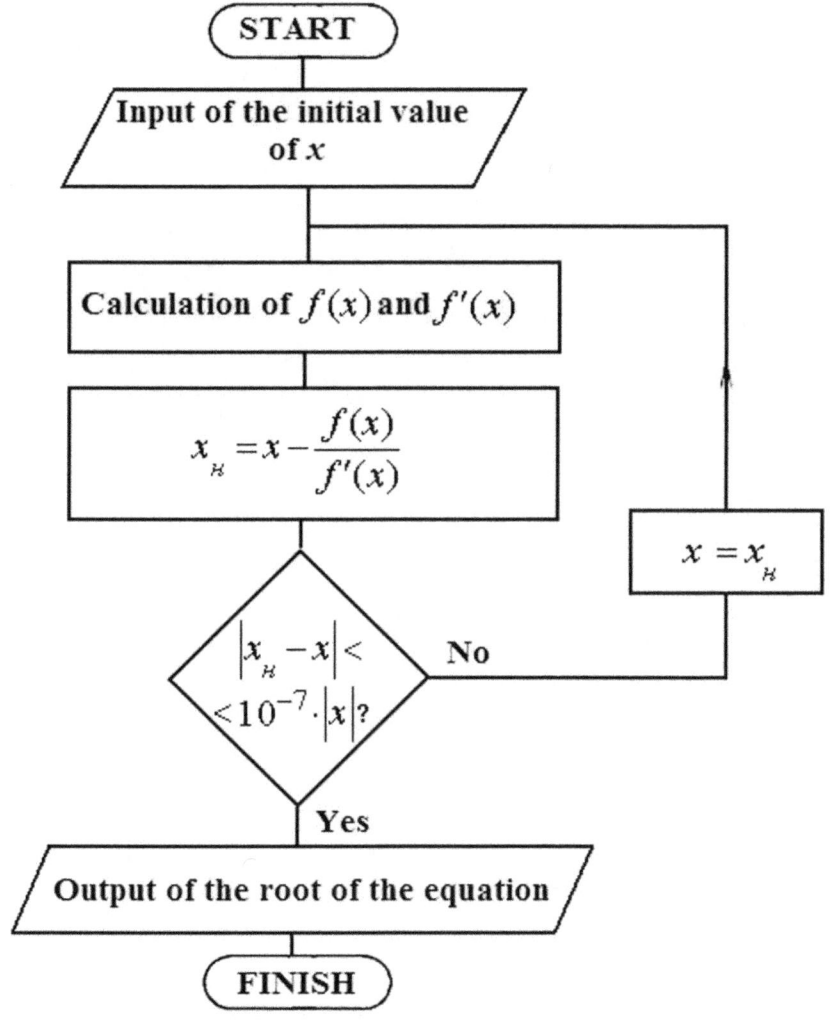

SVETLANA KARITSKAYA

ABOUT THE AUTHOR

Svetlana Karitskaya, Ph.D., is a docent/assistant professor in the Mathematics Department at Ural Federal University named after The First President of Russia Boris Yeltsin. She teaches general mathematics to undergraduate and graduate level students specialized in thermal engineering, electrical engineering and electrical technology, metallurgy, mechanical engineering and materials processing.

Svetlana received her Ph.D. in thermal and molecular physics from Moscow State Forest University in 1997. With respect to her research work, she studies modern computational methods in applied spectroscopy and material characterization. In addition to her teaching and research work, she administers and coordinates conferences for high school students in physics and mathematics.

Svetlana's recent textbooks published in Kazakhstan and Russia:

Karitskaya S.G., Higher Mathematics. Ekaterinburg, Ural Federal University, 2012. - 229 p.

Karitskaya S.G., Numerical Methods and Fundamentals of Computer Simulation.
Ekaterinburg, Ural Federal University, 2011. - 62 p.

Karitskaya S.G., Molecular Physics and Thermodynamics. Laboratory Practice.
Karaganda, Karaganda State University,

SVETLANA KARITSKAYA

* 9 7 8 1 4 9 0 9 3 8 7 0 7 *